UTERINE INFECTION IN COWS

UTERINE INFECTION IN COWS

By:
Pankaj Sood
Department of Veterinary Gynaecology and Obstetrics
DGCN College of Veterinary and Animal Sciences
CSK Himachal Pradesh Agricultural University
Palampur-176 062, Himachal Pradesh

CRC Press
Taylor & Francis Group
Boca Raton London New York

CRC Press is an imprint of the
Taylor & Francis Group, an **informa** business

NPH

NARENDRA PUBLISHING HOUSE
DELHI (INDIA)

First published 2026
by CRC Press
4 Park Square, Milton Park, Abingdon, Oxon, OX14 4RN

and by CRC Press
2385 NW Executive Center Drive, Suite 320, Boca Raton FL 33431

CRC Press is an imprint of Informa UK Limited

British Library Cataloguing-in-Publication Data
A catalogue record for this book is available from the British Library

Print edition not for sale in South Asia (India, Sri Lanka, Nepal, Bangladesh, Pakistan and Bhutan)

ISBN13: 9781041247173 (hbk)
ISBN13: 9781041261452 (pbk)
ISBN13: 9781003749257 (ebk)

DOI: 10.4324/9781003749257

Typeset in Times New Roman
by Amrit Graphics, Shahdara,
Delhi-110032

DEDICATION

To....

THE ALMIGHTY

&

From a disciple....

To his teachers....

Prof. A. K. Srivastava

Prof. A.S. Nanda

CONTENTS

PREFACE

The book "Uterine Infection in Cows" addresses various aspects of uterine infections, including their classification into different forms. The clinical manifestations of different types and grades of uterine infections have been explored. Uterine infections are liable to cause complications for varying durations in cows, which if not correctly diagnosed and treated can have long-term implications. The effects of uterine infections extend beyond the level of the uterus or other parts of the reproductive tract and can jeopardize the functioning of the hypothalamus-pituitary-ovarian axis in the affected cows. This emphatically delineates a complete understanding of the subject. The diagnosis of uterine infections can prove to be challenging or even erroneous at times, with the possibility of being frequently under diagnosed in several cows due to its diverse range of potential clinical characteristics. The symptomatic and asymptomatic forms of the disease have been addressed in this book, which may prevent recurrent pregnancy loss and infertility. It is imperative to utilize appropriate diagnostic tools and assessments to screen endometritis in cows. As a result, management demands prompt and precise identification of the ailment and the appropriate therapeutics. The needs of individual patient in terms of symptoms and severity are required to be considered prior to resorting to a treatment from a wide array of options. Taking into consideration the available literature, different aspects of uterine infections have been extensively elaborated. The author is confident that the updated information in this book will be an invaluable resource for research and educational needs. Having more than 1000 cited references, the book will serve as an excellent reference source. Besides, the author is also sure that the information will pique interest among the students to undertake a deep-rooted study on the perspective developments in the area.

The author is deeply indebted to the late Professor J. M. Nigam and the late Dr. G.C. Negi, the pioneers of the College of Veterinary and Animal Sciences, Palampur, for providing an academic and professional platform for countless individuals like myself.

Professor (Dr.) Pankaj Sood
B.V.Sc & A.H, M.V.Sc (Gold Medalist),
Ph.D, Post-Doctorate (Israel)
Head, Department of Veterinary Gynaecology and Obstetrics

ABOUT THE AUTHOR

Dr. Pankaj Sood, the author, is a 1992 graduate from the D. G C Negi College of Veterinary and Animal Sciences, Himachal Pradesh Agricultural University, Palampur. Subsequently, the author completed his M.V.Sc., with a Gold Medal, and Ph.D. (as an in- service candidate), respectively, in 1995 and 2005, in the discipline of Veterinary Gynaecology and Obstetrics from Punjab Agricultural University, Ludhiana, Punjab, India. As a process of academic pursuit, the author undertook post-doctoral studies on cow infertility in Israel during 2013-14. Dr. Sood was an ICAR-JRF topper (Veterinary Gynaecology and Obstetrics) and has several other fellowships, scholarships and international travel grants to his credit. The author has worked in teaching, research, public outreach and administration for over 29 years. Currently, he is a Professor and Head of the Department of Veterinary Gynaecolgy and Obstetrics in Palampur.

CHAPTER 1

EFFECT OF UTERINE INFECTIONS ON CERTAIN CLINICO-GYNAECOLOGICAL PARAMETERS IN COWS

Pankaj Sood and Vijender Negi

Uterine diseases in cattle occur at all stages of the reproduction cycle but the majority of cases are found in the postpartum period. Uterine infection usually arises as part of a complex of diseases that is referred to as the 'metritis complex' and may involve different investments of uterus and surrounding structures, all of which can occur in varying severities. Nevertheless, the different forms of infection share the same causes, often lead to one another and have treatments in common; so it makes sense to consider them together. The balance between pathogenicity (disease causing ability) and immunity of the host animal decides the setting up and course of infection. The latter is affected by several host level factors, for instance negative energy balance. Hence, a deranged inflammatory process in the absence of a concurrent infection has also been cited as a reason for subclinical forms of uterine infection (Mattmiller *et al.* 2013).

Several fertility parameters have been used to describe negative effects of uterine diseases on reproductive performance, e.g. days to first service, estrus detection rate, and probability of pregnancy per insemination. Days to first service and estrus detection rate are mainly affected by management (time for estrus detection, accuracy of estrus detection, claw trimming, and several other aspects related to management) and to lesser extent by cows' diseases, e.g. cysts, conception risk is strongly related to uterine health. Thus, the prevention of uterine diseases, efficient detection of affected cows and efficacious treatment strategies are important aspects of the fertility management of dairy cows.

Whether uterine diseases have a negligible, moderate or detrimental effect on fertility is still under discussion and depends on definitions and classification.

DOI: 10.4324/9781003749257-1

Irrespective of the underlying etiology, uterine infection damages the uterus and disturbs the uterine microenvironment affecting fertilization and/or embryonic survival thereby affecting the reproductive performance (de Boer *et al.* 2015a). The effects of uterine infections extend much beyond the uterus. Bacterial endotoxins have numerous effects on hypothalamo-pituitary-utero-ovarian axis, which includes reduced luteinizing hormone production (Lavon *et al.* 2010), imbalanced PGF2α-PGE2 secretion (Herath *et al.* 2009), altered ovarian steroid production, ovulation failure (Lavon *et al.* 2008) as well as abnormal follicular and luteal growth (Herath *et al.* 2007; Shimizu *et al.* 2012). In a more recent study, clinical endometritis contributed towards atypical ovarian profiles, which included prolonged luteal activity as the most common implication; the others in descending order being cessation of cyclicity, short luteal phase and delayed first ovulation in postpartum cows (Mohammed *et al.* 2019). Interestingly, the intensity of the undesirable effects on different reproductive segments as well as the carryover effects of the infection also vary according to the gram staining attributes of the microbes. For instance, gram-negative compared to gram-positive bacteria are more resistant (Breijyeh *et al.* 2020) and more harmful (Moyes *et al.* 2009). Moreover, compared to gram-positive, gram-negative bacteria, such as *Escherichia coli,* triggers an immediate and more severe effect, whereas gram-positive bacteria, such as *Staphyloccocus aureus,* has both immediate and long-term effect in disrupting structural and steroidogenic function of the ovarian follicles (Lavon *et al.* 2011). Besides gram staining characteristics, an increase in the pathogen growth density proportionately increased the magnitude of reproductive perturbations (Williams *et al.* 2007). Hence, it is plausible to anticipate multilevel undesirable effects on reproduction in the cows with higher grades of uterine infections, such as metritis and clinical endometritis involving recognized pathogens.

It is noteworthy, that majority of the studies on uterine infections *vis-à-vis* reproduction confine to the fresh postpartum cows having distinctively different metabolic, immunological, reproductive and physiological status (Sheldon *et al.* 2002b; Williams *et al.* 2007; Sheldon *et al.* 2019), which needs to be put in perspective when compared to regular cycling cows. These, along with a discrepancy in the microbial involvement as well as their extended effects on the hypothalamo-pituitary-ovarian axis in the cows with subclinical endometritis have been addressed to varying extent.

A. REPRODUCTIVE DERANGEMENT DUE TO UTERINE INFECTIONS

a. Uterine Tone

The uterine tone varies with the stage of estrous cycle in cows and may be under a common hormonal control. The estradiol increases the contractility as well as intercellular accumulation of water between the uterine myofibrils, resulting in an overall increase in the size and turgidity of the uterine horns (Hafez and Hafez, 2000). Uterine horns become more tonus with an increase in the circulatory estradiol concentrations and contractility reached a plateau before the estradiol peak and sustained until ovulation (Sumiyoshi *et al.* 2014). Hence, it is reasonable to assume that the cows having higher preovualtory estradiol concentrations would have a greater uterine tone (Bonafos *et al.* 1995; Hafez and Hafez, 2000). Amri *et al.* (2015) adopting timed AI in cows induced for ovulation found uterine tone to be a useful indicator for the success of AI to pregnancy.

Transrectal confirmation of the presence of uterine tone and a clear genital discharge may be vital for evaluating and breeding of the cows using AI (Sood *et al.* 2009; Lambertz *et al.* 2014). Good uterine tone along with clear, copious and stringy discharge have been collectively used for confirmation of estrus / ruling out mid – cyclic estrus in cows presented for insemination (Sood *et al.* 2009).

b. Genital Discharge

Upon introduction into the female genital tract, the sperms are immediately suspended into the genital secretions. The genital discharge is therefore a vital component of fertility as it affects the activity, viability and transport of sperms (Tsiligianni *et al.* 2011; Layek *et al.* 2013). The rheological and biochemical properties of the genital discharge undergo a discrete alteration during different phases of the estrous cycle, primarily due to variation in the ovarian steroid production (Blair *et al.* 1941; Hamana *et al.* 1971; Murugavel and Lopez-Gatius, 2009). During estrus, the genital discharge increases in quantity, becomes transparent/clear and flows readily. During luteal phase, the discharge becomes scanty, thick, and opaque and often shows floccules due to increased presence of leucocytes, as also presence of blood in some normal cows (Elstein, 1974; Hafez and Hafez, 2000; Noakes *et al.* 2019). The appearance of genital discharge indicates the apparent health of the reproductive tract and its visual appearance during estrus is used as an instant indication of genital infection (Gill *et al.* 1973; Tsiligianni *et al.* 2011). For instance, genital discharge exhibiting turbidity, flecks of white of off-white mucopurulent material is considered as a clear-cut indication of the underlying genital infection (Williams *et al.* 2005). To the contrary, presence of turbidity in the discharge may also be a reflection of physiological elimination of genital microbiome with otherwise healthy uterus is some cows (Roberts, 1986). Apart from appearance, the consistency and spinnbarkeit properties of the genital discharge are equally important in assessment of the health of the reproductive tract prior to insemination. The latter assumes a greater importance in subclinical infections of genital tract, whereby the genital discharge appears as good as normal (Gilbert *et al.* 2005; Sheldon *et al.* 2006). It is known that *Trueperella pyogenes* binds to the fimbrial protein subunits of the mucosal secretions of the female genital tract. Upon attachment, *Trueperella pyogenes* releases neuraminidase (nan), such as nanH and nanP, which splits the sialic acid component of the genital discharge and reduces the viscosity of the discharge (Jost *et al.* 2001, 2002; Esmay *et al.* 2003; Pietrocola *et al.* 2007).

On the basis of physical appearance of the genital discharge in postpartum cows, it has been graded from 0 (clear and translucent) to 3 (discharge having ≥ 50% pus) (Sheldon, 2004). A higher discharge score generally indicates infection of higher severity (Sheldon 2004; Noakes *et al.* 2019) as well as a relatively longer time required for clinical recoveries in cows (Sheldon and Noakes, 1998; Sheldon *et al.* 2009a).

The cervical mucus was transparent in 58.3%, turbid in 31.5% and dirty in 10.2% normal dairy cows. Further, the mucus consistency was thin in 74.1% and thick in 25.9% cows. In terms of pregnancy outcome on the basis of appearance and consistency of genital discharge, the findings revealed that 67.3% 23.6% and 9.1% mucus samples were transparent, turbid and dirty in the cows getting pregnant. The corresponding figures for the non-pregnant cows were 49.1%, 39.6% and 11.3%, respectively. On the other hand, the pregnancy rates in cows with thin and thick consistency of cervical mucus were 81.8% and 18.2%, respectively. It was inferred that a clear discharge of thin consistency was conducive for pregnancy (Lim *et al.* 2014). Similar findings were reiterated by Bernardi *et al.* (2016), who recorded transparent and filant (between watery and viscous) discharge resembling egg white in 65% of the cows getting pregnant; thick mucus on the other hand was associated with 60% non-pregnant cows. Enkhia and Kohli (1982) recorded transparent mucus in 50% repeat breeder cows, whereas it was translucent in 30% and yellowish in remaining 20% cows. In another study, the mucus in repeat breeder cows had higher (66.7%) turbidity levels than that of normal cows having clear (100%) mucus. The presence of thick, moderate and thin consistencies of cervical mucus in fertile cattle was 0.00, 33.3, and 66.7%, respectively. All (100%) repeat breeder cattle had thick mucus (Siregar *et al.*

2019). According to Mufti *et al.* (2010), 47.5% conception was observed in cows having the mucus consistency ranging from thin to moderate, that in turn was believed to increase the migration of spermatozoa inside female reproductive organs. Combination between mucoprotein and weak spermatozoa penetration might suppress progressive movement of spermatozoa in cows having thick cervical mucus and therefore reduced conception rates.

c. Estrus Duration

The estrus duration averages 12 to 18 hours, with a range of 2 to 50 hours in normal cows under most conditions (Hopper, 2014). Few previous studies have observed longer estrus duration in the repeat breeder compared to the control cows. In one such study, estrus duration was marginally longer in repeat breeders compared to normal cows, the respective values being 21.4 h and 19.6 h (Sood *et al.* 2015). In another study by Bage *et al.* (2002), estrus duration was significantly prolonged in repeat breeder (20.9 h) compared to control (15.2 h) heifers. Bage *et al.* (2002) attributed prolonged estrus in repeat breeding to peri-ovulatory suprabasal progesterone (progesterone concentrations above basal concentrations of 0.5 ng/mL during estrus). Suprabasal progesterone at estrus results in absence or delayed emergence of distinct preovulatory luteinizing hormone peak and therefore a corresponding prolongation of estrus duration in experimental (Duchens *et al.* 1994;1995) and spontaneous repeat breeder (Gustafsson *et al.* 1986; Albihn *et al.* 1991) cows. Moreover, repeated inseminations performed until ovulation, detected ultrasonographically, in prolonged estrus cows did not yield pregnancy (Duchens *et al.* 1995). In contrast to the findings of Bage *et al.* (2002), the timing and release of luteinizing hormone surge were unaffected and instead reduced pre-ovulatory luteinizing hormone concentrations were observed in repeat breeder cows (Sood *et al.* 2015). It is pertinent to mention that Bage *et al.* (2002) did not rule out a possibility of presence of uterine infection in spontaneous prolonged estrus repeat breeding cows. Similarly, Sood *et al.* (2015) neither evaluated subclinical endometritis nor suprabasal progesterone in repeat breeder cows that were given multiple injections of PGF2α for experimental induction of estrus during the investigation course.

Experimental inoculation of endotoxins (simulating the infection like situation) reduced or delayed the luteinizing hormone surge and prolonged the estrus duration from 30 h to 75 h in one-third of the experimental cows (Lavon *et al.* 2008). More recently, a tendency of prolonged estrus in spontaneous cycling cows with concomitant clinical endometritis was also observed (Sood and Ahuja, 2023).

d. Estrus Onset to Ovulation Interval

The average estrus onset to ovulation interval ranges from 27 to 30 h (range of 24 to 38 h) (Larsson, 1987; Hafez and Hafez, 2000; Roelofs *et al.* 2005; Youngquist and Threlfall, 2006; Skidmore, 2015; Noakes *et al.* 2019). Repeat breeding cows with one or the other endocrine disturbance, consistently exhibited a lengthier estrus onset to ovulation interval. The values in lactating cows *versus* heifers was 38.5 h *versus* 29.4 h (Saumande and Humblot, 2005), whereas that in repeat breeders *versus* normal cows were 28.4 h *versus* 24.1 h (Bage *et al.* 2002) and 31.5 *versus* 28.9 h (Sood *et al.* 2015), respectively. Experimental administration of endotoxin at the onset of estrus extended the estrus to ovulation interval to 75 h in one-third of the cow population and was significantly longer than 29 h in the endotoxin infused, but unaffected, and 31 h in the control cows (Lavon *et al.* 2008).

Variation in estrus onset to ovulation interval is primarily influenced by the estrus duration. Bloch *et al.* (2006) determined the estrus onset to ovulation interval in different sets of cows having variable estrus durations. Accordingly, the cows with estrus onset to ovulation interval of 23.9 h, 28.2 h, 32.3 h and 42.6 h were classified as the ones with short estrus, normal estrus, long estrus and very long estrus, respectively. Furthermore, the cows with long and very long estrus had altered profile of progesterone, estradiol and luteinizing hormone surge in plasma. Using the classification of Bloch *et al.* (2006), the estrus onset to ovulation interval, cited *vide supra*, in lactating cows with prolonged estrus by Saumande and Humblot (2005) and in repeat breeders by Sood *et al.* (2015) can be considered in the category of long estrus, while the cows affected by endotoxin challenge in the study of Lavon *et al.* (2008) fall in the category of very long estrus. From a practical point of view, the long interval between the onset of estrus and ovulation, along with its high variation, can be a source of low fertility (Bage *et al.* 2002; Saumande and Humblot, 2005; Bloch *et al.* 2006).

e. End of Estrus to Ovulation Interval

In contrast to a wide variation in estrus onset to ovulation interval, the variation in the end of estrus to ovulation interval is relatively less. The average duration of end of estrus to ovulation interval across different studies ranged from 11.6 to 15.0 h in normal cows (Larsson, 1987; Hafez and Hafez, 2000; Roelofs *et al.* 2005; Youngquist and Threlfall, 2006; Skidmore, 2015). No difference in the said interval was observed between repeat breeders (10.1 h) and normal (9.3 h) cows (Sood *et al.* 2015).

f. Length of Estrous Cycle / Interestrus Interval

The average length of the estrous cycle is 17 to 24 days in cows (Sood *et al.* 2022). Amongst different factors, presence of uterine infection can change the length of estrous cycle in cows. The length may increase or decrease in some, while it may be unaltered in the others. The alterations *per se*, if present, have been linked to a change in the length of luteal phase in the affected cows (Opsomer *et al.* 2000; Herath *et al.* 2006; Sood *et al.* 2022).

Uterine infection related factors influencing the length of estrous cycle include the quantum of infection, the qualitative spectrum of infecting microbes and the individual response to infection (Olson *et al.* 1984; Farin *et al.* 1989; Peter *et al.* 1989; Williams *et al.* 2007; Kaneko and Kawakami, 2008). It is believed that the foremost tendency of uterine infection is to lyse the corpus luteum; another response to uterine infection may be an extension of the luteal phase and therefore increase the length of the cycle (Galvao, 2018). Numerous studies have reported an increase in the length of estrous cycles in cows with heavily contaminated uterus (Olson *et al.* 1984,1986; Opsomer *et al.* 2000), including closed pyometra (Etherington *et al.* 1991). In agreement to the latter studies, a subsequent report investigating variable quantum of infection *vis-à-vis* length of luteal phase, observed a higher percentage of cows with mild uterine infection to exhibit a short luteal phase, whereas a long luteal phase was a consistent feature in the cows with severe uterine infection (Mateus *et al.* 2002a). In terms of the effect of qualitative characteristics of microbes on estrous cycle length, uterine infection from *Escherichia coli* (Peter *et al.* 1989; Mann and Lamming, 1995; Mateus *et al.* 2002a; Williams *et al.* 2007; Herath *et al.* 2009) and *Trueperella pyogenes* (Farin *et al.* 1989) increased the length of estrous cycle in cows (Mann and Lamming, 1995; Williams *et al.* 2007; Herath *et al.* 2009). Individual sensitivity as a variant for interestrus interval in light of the underlying infection was substantiated by Kaneko and Kawakami (2008). Upon a single experimental inoculation of uterus with a similar microbial load of *Trueperella pyogenes* three days after spontaneous ovulation (diestrus) in a set of cyclic cows, 50 percent of the cows experienced a normal life span of corpus luteum, whereas the corpus luteum regressed prematurely in the remaining cows. Interestingly, the PGFM concentrations were higher in the unaffected and affected cows exhibiting a normal and reduced life span of corpus luteum, respectively. It was also noteworthy that the cows instilled with *Trueperella pyogenes* exuded a clear or translucent genital discharge, which rules out a heavy infection, as discharge of similar appearance is present during diestrus in normal non-infected cows (Sood *et al.* 2009). As a process of inflammation reaction to infection, uterine tissue, like other body tissues, can synthesize different types of prostaglandins, namely, PGF2α and PGE2 (Guilbault *et al.* 1984a). PGE_2 and $PGF_2α$ have opposite effects on the existing corpus luteum. While PGE_2 has antiluteolytic and/or luteotrophic properties (Magness *et al.* 1981), PGF2α is strictly luteolytic (Asselin *et al.* 1996; McCracken *et al.* 1999) in cows. It is also known that the epithelial cells of endometrium exposed to bacterial endotoxin undergo an endocrine switch from luteolytic PGF2α to luteotrophic PGE2 production (Herath *et al.* 2009). Hence, a variation in the PGF2α: PGE2 ratio may determine the fate of corpus luteum and the interestrus interval in cows ridden with uterine infection. Variation in the type of inflammatory mediators produced during infection also determines the type of PG synthesis. For instance, inflammatory cytokines such as TNF alpha stimulates the release of PGF2α from the endometrium and luteal cells (Kaneko and Kawakami, 2008, 2009), whereas, IL-1 and IL-6 decrease the expression of oxytocin receptors in endometrial cells, which could impair or delay the mechanism of leuteolysis (Leung *et al.* 2001). Hence, inflammation has a bimodal effect on the length of estrous cycle. The qualitative characteristics of microbial infection also dictate the PGE2: PGF2α ratio. *In vitro* studies on tissue explants (Herath *et al.* 2009) and *in situ* uterine infection (Mann and Lamming, 1995; Mateus *et al.* 2002a; Williams *et al.* 2007) have demonstrated *Escherichia coli* to increase PGE2 to PGF2α ratio in cows.

Furthermore, presence of a persistent corpus luteum in 33% of the metritic cows which did not differ from the healthy subjects in terms of PGE2 concentrations during early postpartum period (Struve *et al.* 2013), supports the likelihood of individual variation.

g. Ovulatory Follicle

The average size of the ovulatory follicle ranged from 11.0 to 15.0 mm in Holstein heifers, which was smaller than 13.1 to 17.2 mm in the lactating Holstein cows (Sartori *et al.* 2004; Noakes *et al.* 2019). The average maximum size of the ovulatory follicle in Jersey crossbreds ranged from 12.1 to 12.9 mm (Sood *et al.* 2022). The size of the ovulatory follicle significantly influenced the fixed time AI pregnancy rates in cows receiving GnRH for induction of ovulation (Perry *et al.* 2007). On the other hand, there was no effect of follicular size on pregnancy outcome in cows undergoing spontaneous ovulations (Perry *et al.* 2005). The latter two studies as well as several others accumulate convincing evidence that it is the maturity of the ovulatory follicle rather than its size to influence the establishment and maintenance of pregnancy (Bridges *et al.* 2010; Geary *et al.* 2013; Dickinson, 2016). The measure of maturity of the ovulatory follicle is its ability to produce greater amounts of estradiol prior to ovulation and its subsequent differentiation into a corpus luteum with a capacity of producing adequate amounts of progesterone conducive for pregnancy. Cows that had relatively greater plasma estradiol concentrations also had greater plasma progesterone concentrations and higher pregnancy rates (Vasconcelos *et al.* 2001; Atkins *et al.* 2013). Lower plasma estradiol concentrations preceding and plasma progesterone following ovulation may result in inadequate preparation of the maternal tissues for pregnancy (Pohler *et al.* 2012; Dickinson, 2016). It is also logical that the oocyte from a more mature ovulatory follicle would have a better embryonic developmental capacity (Perry *et al.* 2005).

The size of the ovulatory follicle varies between and within breeds (Sartori and Barros, 2011) as well as due to parity in cows. Reduced body condition score lowers the functional capacity of the ovulatory follicle and yields reduced pregnancy rates (Ciernia *et al.* 2021). Presence of uterine infections also alters the follicular characteristics. Cows with genital tract microbial infections had reduced growth rate of the dominant follicles, lower plasma estradiol around the time of maximal follicular diameter as well as progesterone concentration and were less likely to ovulate. The undesirable follicular characteristics were more marked in cows with heavy rather than low uterine pathogen growth density at day 7 postpartum (Sheldon *et al.* 2002b; Williams *et al.* 2007). The slower growth rate of the dominant follicle and its reduced ability to ovulate were preferably associated with the presence of *Escherichia coli* and *Trueperella pyogenes* in the uterine lumen (Williams *et al.* 2007). Further, the lipopolysaccharide presence in the follicle may affect the quality microenvironment milieu for the final stages of oocyte maturation in repeat breeder dairy cows with subclinical endometritis (Heidari *et al.* 2019).

The granulosa cells have the ability to generate an immune reaction and secrete inflammatory cytokines in response to infection/bacterial products entering from circulation. Due to immune reaction, the primary function of steroidogenesis in the granulosa cells gets compromised (Dahiya *et al.* 2018).

h. Endocrinological Perturbations

(i) **Corpus luteum and progesterone:** The luteinizing hormone is essential for maintenance of progesterone production from the luteal cells in cattle. The luteinizing hormone pulses and progesterone production are not associated during the early luteal phase; however, the luteinizing hormone pulses are followed by progesterone release during mid-luteal phase (Quintal-Franco *et al.* 1999). Nonetheless, the preovulatory to ovulatory phase luteinizing hormone surge concentrations are vital in regulating the functional health of the corpus luteum during luteal phase. Low luteinizing hormone surge concentrations or its delay reduces the luteal phase production of progesterone from corpus luteum (Gustafsson *et al.* 1986; Albihn *et al.* 1991). Suprabasal progesterone is one potential reason for a delayed or reduced luteinizing hormone surge in spontaneous repeat breeder cows (Duchens *et al.* 1994;1995).

The diameter of the corpus luteum may provide information on the functionality of the corpus luteum. The size of the corpus luteum, though variable among different studies, tends to increase till day 5 to 7 post ovulation and in consonance so does the progesterone, both of which fluctuate between day 7 to 13 and start reducing by day 17 or 19 onwards (Quintal-Franco *et al.* 1999; Noakes *et al.* 2019). The latter depends on the length of the luteal phase in cows, which differs due to varying follicular wave patterns (Sood *et al.* 2022). The size of the corpus luteum and the corresponding progesterone concentrations between day 3 to 7 of estrous cycle varied from 15.0 to 23.0 mm and 7.0 to 13.0 ng/mL in cows and heifers, respectively (Assey *et al.* 1993). In another study on cows having different estrous cycle lengths of 13 days (short cycle), 22 days (normal cycle) and long estrous cycle (30 days), the rise in plasma progesterone concentration was slow in the cows with short cycle as compared to the ones with normal and long cycles. Moreover, without any difference in the maximum size of the corpus luteum between all the groups, the peak progesterone concentrations was acquired at day 8, 10 and 18 in cows with short, normal and long estrous cycles, respectively (Kayacik *et al.* 2005). The pattern of progesterone secretion during early luteal phase is important in deciding successful pregnancy outcome. The rate of corpus luteum development between day 5 and 10 after insemination is considered as an accurate indicator of corpus luteum function, and a predictor of the ensuing pregnancy. Low progesterone concentration during early luteal phase was recorded in non-pregnant compared to pregnant buffalo-cows (Mansour *et al.* 2016). Starbuck (1999) framed a benchmark progesterone concentration at day 5 after insemination *vis-à-vis* pregnancy. Accordingly, the pregnancy was unlikely in cows having plasma progesterone concentration of <3 ng/mL on day 5 after AI. Few other studies also agree that blood progesterone concentration in pregnant cows diverge from those that fail to conceive at some point between day 6 and 14 post insemination and that this divergence affects the *in utero* embryonic development (Bulman and Lamming, 1978; Lamming, 1989).

Kendall *et al.* (2009) observed that low progesterone concentration (<3ng / mL) on day 5 was more common in repeat breeders (34%) (not assigned any cause) than in normal cows (11%). In another study, the progesterone concentration in repeat breeder heifers with prolonged estrus, at day 5 was 5 ng/mL, which was less compared to 7 ng/mL in the normal heifers at the same stage of cyclicity; moreover, a similar discrepancy between the two groups was maintained at day 3, 7, 13 and 17 of the estrous cycle (Bage *et al.* 2002).

Uterine infections and their inflammatory products postpartum could alter the luteal activity as well as lead to follicular dysfunctions such as smaller follicles and reduced estradiol concentrations. Prolonged luteal phase has been the most common form of luteal abnormalities and was recorded in 32 to 60% of the cows with postpartum uterine infections (Taylor *et al.* 2003; Ranasinghe *et al.* 2011; Ghanem *et al.* 2015; Mohammed *et al.* 2019). Using vaginal discharge to classify the severity of postpartum uterine infection, prolonged luteal activity was observed to be significantly correlated with severity of infection in some studies (Williams *et al.* 2007; Gorzecka *et al.* 2011). Another study, however, failed to detect any such relationship (McCoy *et al.* 2006). A greater degree of uterine infections brings undesirable ovarian effects in a greater percentage of infected cows as also an increased intensity of ovarian perturbations (Mohammed *et al.* 2019). The latter finding is justified by a greater absorption of endotoxins from uterus into systemic circulation which was detected in blood plasma of all the heavily infected cows (n=8; 100%) compared to the endotoxin detection in one out of six (17%) mildly infected cows (Mateus *et al.* 2002a,b). Another point of interest in the present context would the effect of uterine infections on the size of corpus luteum and progesterone production. Struve et al. (2013) investigated corpus luteum and progesterone for four consecutive postpartum estrous cycles in metritis cows. Presence of metritis reduced the size of corpus luteum, but had no effect on progesterone concentrations during the first cycle. However, the size of corpus luteum as well as progesterone remained unaffected during the subsequent cycles in the infected cows. Hence, metritis impaired luteal activity transiently without any long-term effect. On the basis of uterine pathogen growth density at day 7 postpartum, Williams *et al.* (2007) found that the size of the first postpartum corpus luteum was not markedly different, but the peripheral plasma progesterone concentration from first postpartum corpus luteum was reduced in the cows with high (<2 ng/mL) *versus* low (5ng/mL) pathogen growth density. The latter phenomenon is in consonance to a previous study, where bovine viral diarrhea virus infection resulted in a suppression of progesterone concentrations (Fray *et al.* 1999). Experimental inoculation of *Escherichia coli* endotoxin through intravenous or intrauterine route in cyclic cows transiently reduced the size of corpus luteum and circulating progesterone; the cycle length, however, remained unaffected (Gilbert *et al.* 1990; Williams *et al.* 2008; Herzog *et al.* 2019). Uterine infections, either spontaneous or induced, have a potential to delay or block the luteinizing hormone surge and suppress the luteinizing hormone secretion (Lavon *et al.* 2008). The inflammatory mediators produced at the uterine level may significantly reduce the concentration of luteinizing hormone receptors in the follicle, besides having a direct effect on the hypothalamus-pituitary axis to reduce the circulatory luteinizing hormone and therefore progesterone; all this compromises the pregnancy outcome in the infected cows (Karsch *et al.* 2002).

(ii) Preovulatory estradiol surge and concentrations: Typically, gradualy increasing estradiol concentrations attain a peak 12 to 18 h before the onset of estrus that triggers a preovulatory surge of GnRH and luteinizing hormone in the normal cows (Coe and Allrich, 1989; Cook *et al.* 1991; Xu *et al.* 1997; Yaniz *et al.* 2006; Reames *et al.* 2011; Hooper, 2014). In contrast, several other studies reported estradiol surge to coincide with estrus (Dieleman *et al.* 1986) or may even occur 3.1 (Stevenson *et al.* 1998) to 6.8 h (Saumande and Humblot, 2005) after estrus. In agreement to the latter set of studies, the estradiol peak was recorded 2.5 h and 2.1 h after estrus, respectively, in repeat breeder and normal cows (Sood *et al.* 2015).

The peak estradiol concentration ranges from 7 to 47 pg/mL. A high variation *per se* is attributed to breed, parity, milk production, peri-estrual stage, spontaneous *versus* induced estrus and method of estimation (Coe and Allrich, 1989; Cook *et al.* 1991; Jinks *et al.* 2013; Bernardi *et al.* 2016; Noakes *et al.* 2019). In another study, the preovulatory estradiol concentration (pg/mL) was 2.5, 3.1, 3.5 and 6.5 at 6, 12, 18 and 24 h prior to ovulation (Sumiyoshi *et al.* 2014). In normal *versus* repeat breeder cows of nearly similar estrus duration, the peak estradiol concentrations (pg/mL) were 18.8 *versus* 22.7, respectively. At 12, 18 and 24 h after estrus onset, the estradiol concentrations (pg/mL) in normal *versus* repeat breeder cows, respectively, were 8, 12 and 18 *versus* 9.5, 14 and 16 (Sood *et al.* 2015). On the other hand, cows experiencing a delayed estrus onset to ovulation interval (prolonged estrus) had a clear evidence of reduced estradiol concentrations. For instance, estradiol surge concentration was significantly less in cows with a e" 35 h estrus onset to ovulation interval (Bloch *et al.* 2006). Also, the prolonged estrus cows with clinical and subclinical mastitis had a 50% reduction in estradiol concentration values at estrus (3 *versus* 6 pg/mL) (Lavon *et al.* 2010). Another study in fresh postpartum cows having high *versus* low uterine pathogen growth density revealed that the daily estradiol concentrations were consistently low from day 10 to 16 in the cows with high pathogen growth density (Williams *et al.* 2007). It is known that the uterine toxins and inflammatory mediators, probably through systemic circulation, enter into the ovarian follicle (Nakajima *et al.* 1997; Blum *et al.* 2000; Hisaeda *et al.* 2001; Sheldon *et al.* 2002b; Herath *et al.* 2007). The ovarian granulosa cells, like other immune cells in the body, possess toll – like receptors that interact with bacterial pathogen associated molecule pattern. During the process, steroidogenesis, the primary function of granulosa cells, is compromised through down regulation of CYP19A1 gene expression culminating into low estradiol production from the affected follicle (Dahiya *et al.* 2018).

Estradiol concentrations preceding ovulation dictates the timing and quantum of luteinizing hormone surge (Saumande and Humblot, 2005; Bloch *et al.* 2006). The other important preovulatory functions of estradiol include its direct role in the regulation of the biological clock in the uterus (Nakamura *et al.* 2005) as also facilitation of sperm transport (Hawk, 1983), regulation of uterine pH (Perry and Perry, 2008), induction of endometrial progesterone receptors (Zelinski *et al.* 1980) and histotroph production (Miller, 1977) in cattle.

(iii) **Preovulatory luteinizing hormone surge and concentrations:** Typically, the chronology of the different events leading to ovulation comprise of a peak estradiol surge, then the luteinizing hormone surge and ovulation (Stevenson *et al.* 1998). Also, there is a high correlation between the drop in estradiol concentration and the occurrence of the preovulatory luteinizing hormone surge, which in turn is correlated with the moment of ovulation (Lyimo *et al.* 2000). Hence, it will not be surprising to expect that all the cows experiencing normal estrus duration and estradiol surge prior to estrus onset, will exhibit luteinizing hormone surge around estrus onset (Bloch *et al.* 2006). Infact, the first signs of estrus coincide with the luteinizing hormone surge in cows (Hopper, 2014). Other cows experiencing estradiol surge at estrus or thereafter would display luteinizing hormone surge in the subsequent part of estrus; the period may range from 2.2 to 3.9 h in cows with normal estrus duration (Bloch *et al.* 2006). Several other reports suggest that nearly 10 to 20 % of the cows and heifers exhibited a delay in luteinizing hormone surge in relation to estrus (Mikeska and Williams, 1988). The repeat breeder animals with prolonged estrus duration in particular had longer intervals from onset of estrus to luteinizing hormone peak than the normal cows (Gustafsson *et al.* 1986; Bage *et al.* 2002). The average estrus onset to luteinizing hormone peak interval in repeat breeder *versus* normal cows was 12.2 h *versus* 4.8 h (Gustafsson *et al.* 1986) and 4.6 *versus* 0.4 h (Bage *et al.* 2002). Reduced preovulatory estradiol concentration was held accountable for a delay in luteinizing hormone surge in cows with prolonged estrus (Bloch *et al.* 2006). To the contrary, repeat breeders and normal cows with similar estrus duration did not differ in the timing of luteinizing hormone surge, which was evident between 9.5 and 9.3 h, respectively after estrus onset and nearly 7.2 h after estradiol surge in both groups of cows (Sood *et al.* 2015). The elevated luteinizing hormone surge generally sustains for 6-12 h (Hafez and Hafez, 2000).

The luteinizing hormone surge concentrations (ng/mL) ranged from 12.6 to 15.3 in normal cows (Layek *et al.* 2013), whereas the concentration at 24, 18, 12 and 6 h prior to ovulation were 7.0, 6.8, 2.1 and 1.8, respectively (Sumiyoshi *et al.* 2014). The surge concentrations in repeat breeders with extended estrus duration (32.4 ± 3.7 *versus* 26.6 ± 1.8 ng/mL) were significantly less (28.0 ng/mL) than the normal (40.7 ng/mL) (Gustafsson *et al.* 1986) cows. However, Sood *et al.* (2015) observed nearly identical luteinizing hormone concentrations of 10.1 and 10.5 ng/mL in repeat breeding and normal cows, respectively, having normal estrus duration. In another study, three different categories of cows with estrus to ovulation interval ranging from 22 to 35 h had luteinizing hormone concentrations of 9.5 to 10.0 ng/mL, which were significantly higher than 4.0 ng/mL in cows with an estrus to ovulation interval exceeding 36 h (Bloch *et al.* 2006). Presence of endotoxins either delays or prevents the luteinizing hormone surge in cows. Lavon *et al.* (2008) demonstrated that an intravenous or intramammary endotoxin exposure at peak estradiol concentrations reduced the luteinizing hormone surge concentration in some cows to 3.2 ng/mL, which was significantly less than 14 ng/mL in the untreated cows at same reproductive stage. The timing of luteinizing hormone surge was, however, unaffected. The inflammatory mediators associated with uterine infection may significantly reduce the follicular concentration of luteinizing hormone receptors, besides having a direct effect on GnRH/luteinizing hormone release at hypothalamus-pituitary level (Karsch *et al.* 2002).

CHAPTER 2

DIAGNOSTIC TECHNIQUES FOR UTERINE INFECTIONS IN COWS

Pankaj Sood

Uterine infections have been defined on the basis of the segment of reproductive tract involved or the physical appearance of the genital discharge and the associated systemic signs.

A. DEFINITIONS BASED ON INFLAMMATION OF DIFFERENT SEGMENTS OF REPRODUCTIVE TRACT

a) Endometritis: The inflammation limited to the endometrium.

b) Metritis: Inflammation of the entire thickness of the uterine wall.

c) Perimetritis: Inflammation of the serosa.

d) Parametritis: Inflammation of the suspensory ligaments.

e) Pyometra: Is defined by the accumulation of purulent material within the uterine lumen along with the presence of a persistent corpus luteum and a closed cervix. Although there is functional closure of the cervix, the lumen is not always completely occluded and some pus may discharge through the cervix into the vaginal lumen.

B. DEFINITIONS BASED ON CLINICAL SIGNS AND SYMPTOMS

a. Puerperal Metritis

It is an acute systemic illness due to infection of the uterus with bacteria, which usually and typically occurs within 10 (Sheldon *et al.* 2006) days after parturition and is often a sequel to retained placenta, dystocia, stillbirth or twins. Puerperal metritis, however, may even occur without these conditions and certainly the

DOI: 10.4324/9781003749257-2

presence of membranes is not the cause of the metritis. The disease may even be observed up to 21 days after calving (Sheldon *et al.* 2006, 2009a; Azawi, 2008). This condition is rare after the second week post partum; though generally less severe cases of puerperal metritis may be noticed in the later postpartum stage (Markusfeld, 1984; Azawi, 2008; Drillich *et al.* 2001).

Due to its severe detrimental effects on reproductive performance and economics, acute puerperal metritis is one of the most important postpartum diseases in dairy cows (Azawi, 2008; Overton and Fetrow, 2008; McLaughlin *et al.* 2013). The uterus may be enlarged and the clinical signs of uterine infection vary with the virulence of the causative organisms and the presence of factors that predispose to the disease. Lochia is normally expelled from the reproductive tract during the first few weeks after calving and the discharge may persist for up to 30 days if uterine involution is delayed (Youngquist and Shore, 1997). The color of normal discharge ranges from dark brown to red or white and usually should not be considered abnormal unless the fluid is malodorous or other aberrant clinical signs are noted (LeBlanc *et al.* 2002a). In other words, uterine infections during the puerperal period are often characterized by fetid serosanguineous discharge from the reproductive tract (Roberts, 1986; Noakes *et al.* 2002). When puerperal metritis is complicated by septicemia, toxemia, the milk yield is reduced, and the cow may be unable or unwilling to rise. In such a case, the other clinical signs include fever, dullness, depression, partial or complete anorexia, dehydration, reduced dry matter intake and laminitis (Sheldon *et al.* 2004b). In protracted cases, the high temperature initially, can rapidly drop away to become sub-normal. The acute puerperal metritis has also been associated with visceral pain (Stojkov *et al.* 2015), thereby rendering the disease as an animal welfare issue and consequently the necessity of an efficacious therapy.

The definition of acute puerperal metritis framed by Sheldon *et al.* (2006; 2008; 2009a) comprising of systemic illness and associated toxemia (decreased milk yield, dullness or other signs of toxemia, decreased dry matter intake, elevated heart rate, and dehydration), fever ≥ 39.5°C, signs of toxemia occurring within 21 d after parturition and enlarged uterus along with a watery red-brown fluid to viscous off-white purulent uterine discharge, which often has a fetid odor, has been well accepted and has been used widely in different research trials (Dubuc *et al.* 2011; McLaughlin *et al.* 2013; Sannmann *et al.* 2013b; Sheldon *et al.* 2008). When applying this definition, the reported incidence rate of acute puerperal metritis was 20% (Sannmann *et al.* 2012), but might go up to 40% as mentioned in an earlier study (Markusfeld, 1987). Considering an array of clinical signs and symptoms, previously in research and practice, the most utilized diagnostic criteria to detect a case of acute puerperal metritis includes an elevated rectal temperature (≥39.5°C) and fetid, watery, red vaginal discharge. A systematic review regarding diagnostic methods for acute puerperal metritis demonstrated that these two criteria were described in 39 (fever) and 21 (fetid vaginal) of the 48 peer-reviewed research papers, respectively, addressing acute puerperal metritis (Sannmann *et al.* 2013b). Therefore, this definition has to be considered the best available reference standard. Although, the definition is considered to be empirically based, but certain workers believe that it requires refinement (Burfeind *et al.* 2014a) as there is no gold standard available (Sheldon *et al.* 2006). The latter is based on the existing evidence that vaginal discharge and body temperature are related (Palenik *et al.* 2009; Burfeind *et al.* 2014b) and might therefore be useful to differentiate metritic from healthy cows. However, both criteria are also imperfect and could introduce bias. Several factors that affect body temperature in dairy cows have been described (Burfeind *et al.* 2010). For instance, increasing the fever threshold to 40.0°C for heat-stressed cows has been proposed (Burfeind *et al.* 2012). Also, the sensorial assessment of odor of vaginal discharge is subjective and error prone (Sannmann *et al.* 2013a).

b. Metritis

The stage of onset of metritis postpartum may be same as that for acute puerperal metritis. On the basis of the available literature, metritis refers to a lesser severe form of acute puerperal metritis; although this can range from mild (reduction in appetite and milk yield) to very severe and potentially fatal disease. The term metritis should be used for cows that have delayed involution and a fetid discharge, in the absence of detected fever. Indeed, in some cases of severe bacterial infection pyrexia was not detected, even with daily monitoring of rectal temperature and did not depend on the load or species of bacteria (Sheldon *et al.* 2004b,c). Animals that are not ill but have an abnormally enlarged uterus and a purulent uterine discharge detectable in the vagina, within 21 days after parturition, might therefore be classified as having clinical metritis.

The features of physical examination, alike the ones mentioned for acute puerperal metritis cows, include attitude, hydration status, rectal temperature, and palpation of the uterus per rectum to evaluate uterine discharge in the suspected cows. Evaluation of rectal temperature should be performed before palpation per rectum. In one study (Benzaquen *et al.* 2007) it was observed that a high proportion (~60%) of cows did not have fever (>39.4^{0}C) at the time puerperal metritis was diagnosed, indicating that this condition is not always accompanied by a fever. This finding suggests that diagnosis and treatment consideration for puerperal metritis should include the character of the uterine discharge (fetid or not) and the attitude of the cow, besides measurement of rectal temperature. Cows diagnosed with metritis without a fever were just as likely to later develop clinical endometritis as cows with metritis and a fever. This indicates that metritis without a fever might have the same negative effects on fertility as metritis without a fever (Benzaquen *et al.* 2007).

In contrast, diagnosis of clinical metritis in large herds may be more complicated. Measuring the body temperature daily during the first 10 days after calving and performing vaginal examination at least once between Days 2 to 10 may help diagnosing clinical metritis more accurately. Measuring only rectal temperature is not enough because some cows may have no elevated rectal temperature (>39.5°C) and have low grade puerperal metritis (Sheldon and Dobson, 2004b; Benzaquen *et al.* 2007). The rectal measures of body temperature can be influenced by the procedure itself (up to 0.5^0C), type of thermometer (up to 0.3^0C), and the penetration depth into the rectum (up to 0.4°C difference between a penetration depth of 11.5 cm and 6.0 cm in one of the experiments). Differences in rectal temperature before and after defecation are minor (<0.1^0C). These results may indicate that some care is required in generalizing rectal measures of body temperature (Burfeind *et al.* 2010). Daily measurement of rectal temperature and the assessment of vaginal discharge (color, smell and viscosity) may help the diagnostic accuracy of puerperal metritis (Sannman and Heuwieser, 2015).

According to Sheldon *et al.* (2006), clinical metritis can be categorized into three grades:

Grade 1 clinical metritis: Is characterized by an abnormally enlarged uterus and a purulent uterine discharge detectable in the vagina, within 21 days after calving.

Grade 2 clinical metritis or puerperal metritis: Is characterized by a fetid red-brown watery uterine discharge, atonic enlarged uterus and, usually pyrexia (>39.5°C) (Drillich *et al.* 2001; Foldi *et al.* 2006). In severe cases, reduced milk yield, dullness, in appetence or anorexia, elevated heart rate, and apparent dehydration may also be present. In some cases pyrexia even with daily monitoring of rectal temperature could not be detected (Sheldon and Dobson, 2004; Benzaquen *et al.* 2007). However, an enlarged uterus with a thin wall and atonia was usually detected along with a fetid discharge. Puerperal metritis is often associated with retained placenta, dystocia, stillbirth or twins, and usually occurs toward the end of the first week after calving, being rare after the second week after calving (Markusfeld, 1984; Foldi *et al.* 2006; Konyves *et al.* 2009). It is important to emphasize that puerperal metritis may occur in around 10%-15% of cows with spontaneous calving and without retained fetal membranes (Benzaquen *et al.* 2007).

Grade 3 clinical metritis or toxemic metritis: Is characterized by additional signs of toxaemia (such as cold extremities, depression and/or collapse) which has a poor prognosis (Sheldon *et al.* 2006).

Cows diagnosed with metritis (puerperal or clinical) should be evaluated for concurrent metabolic or infectious disease (ketosis, displaced abomasum, mastitis, pneumonia, etc.) since any of these conditions are associated (Curtis *et al.* 1985).

c. Endometritis (whites)

This is by far the commonest disease in the metritis complex. A 10 to 15 % of the cows develop endometritis after calving. The figures vary greatly between farms with the best farms having this with a prevalence of <4% and the worst over 25%. Clinical form of endometritis can be appreciated by the presence of purulent uterine discharge detectable in the vagina 21 days or more post partum, or mucuopurulent discharge detectable in the vagina after 26 days post partum in the absence of systemic signs, or both (Sheldon *et al.* 2002a,2009a). Azawi (2008) mentioned that some cows with endometritis can be easily identified because they have discharge from the vulva with but a high proportion of cows, probably as high as 80%, show no discharge from the vulva. In some cows, a small amount of pus adheres to the tail. This infection doesn't spread into the systemic circulation, so the cows show no signs of illness. The volume of the discharge is variable, but frequently increases at the time of estrus when the cervix dilates and there is copious vaginal mucus. The cow rarely shows any signs of systemic illness, although in a few cases milk yield and appetite may be slightly depressed (Bearden and Fuquay, 1997). It was also proposed that the term "purulent vaginal discharg≥ be used instead of clinical endometritis, because the presence of genital catarrh was not necessarily indicative of endometrial inflammation (Dubuc *et al.* 2010b).

d. Subclinical Endometritis

Subclinical endometritis is one of the most important reproductive impairments in dairy cows (Sheldon *et al.* 2009a,b; Pascottini *et al.* 2015). Kasimanickam *et al.* (2004) found that many clinically normal postpartum cows had subclinical endometritis. It can be defined as the superficial inflammation of the endometrium (no deeper than the stratum spongiosum) (LeBlanc *et al.* 2002a), without visible clinical signs absent on vaginoscopy too, but significantly affecting reproductive performances (Gautam *et al.* 2009; Sheldon *et al.* 2009b). In addition to absence of pus, the consistency of the genital discharge also remains unaltered (Purohit *et al.* 2015). This subtle endometrial inflammation of the uterus, in the absence of purulent material in the vagina (Dubuc *et al.* 2010a), is generally diagnosed by measuring the proportion of neutrophils present in a sample collected by assessment of the superficial endometrium using flushing or more common through endometrial cytology. In addition, the genital discharge may also lack a pathogenomic property (Kasimanickam *et al.* 2004; Gilbert *et al.* 2005; Sheldon *et al.* 2006). The uterus on transrectal palpation reveals no abnormality (Purohit, 2008).

With no apparent uterine or discharge abnormality, the subclinical form of endometritis may be as detrimental to reproduction as are the other severe forms. Evidence is growing that cows with subclinical endometritis have altered embryo quality and endometrial function. Inflammation in the endometrium has been shown to reduce fertilization in single ovulating postpartum dairy cows (Cerri *et al.* 2009). Dairy cows with no detectable polymorohonuclear cells (PMNs) in endometrial cytology had an increased number of transferable embryos when subjected to superstimulation compared with cows with presence of PMNs in endometrial cytology (Drillich *et al.* 2012). The latter is justified by finding that in the cows diagnosed with subclinical endometritis have altered endometrial and embryonic gene expression that might explain the reduced fertility (Hoelker *et al.* 2012) and has been linked to perturbed cell adhesion and immune modulation, which affects the changes in Day 7 embryo gene expression. The changes in endometrial gene expression might be induced by an altered number of immune cells present in the tissue. Nevertheless, embryos from cows with subclinical endometritis had altered pattern of gene expression involving pathways in cell cycle and apoptosis, which might explain a reduction in pregnancy per AI or even increased risk of pregnancy loss (Hoelker *et al.* 2012). It remains to be elucidated whether subclinical endometritis is the causative agent of changes in the endometrial and embryonic gene expression or that the cows that develop such problems have underlying factors that also cause changes in the transcriptome.

C. METHODS OF DIAGNOSIS

Diagnosis of uterine infections has a great importance because it is generally accepted that up to 40% of the dairy cows may have clinical metritis within the first two weeks after calving and infection may persist in 10 to 15% of the animals more than 3 weeks after calving causing clinical or subclinical endometritis (Sheldon and Dobson, 2004).

There is no quick, easy test for endometritis. The best tests are expensive; the quick tests lack accuracy (Laven, 2019). Nonetheless, clinical records of periparturient problems help in identifying the animals at risk of uterine disease but do not provide specific diagnosis (Markusfeld, 1987; McDougall, 2001). For instance, reduced feeding activity during the pre-partum period can be a significant risk factor for developing different forms of puerperal metritis.

Dairies should have a clear standard operating procedure on when to evaluate cows for metritis and how to identify them. Metritis can occur at any time after calving, even after 21 days in milk; however, most of the cases (~95%) occur in the first 14 days in milk with a peak around 5 to 7 days in milk (Williams *et al.* 2005). Because of this concentration in incidence of metritis in the first 14 days in milk, and in an effort to target monitoring of cows, different strategies have been proposed. Without a regular veterinarian around, the dairy farms can devise a 100 day contract for health and reproductive management as has been adopted in few nations abroad. Part of that program includes daily monitoring of fresh cows in the first 10 days in milk. Animals without systemic signs but with an enlarged uterus and a purulent uterine discharge within 21 days in milk may be classified as having clinical metritis (Sheldon *et al.* 2006). In cases with vulvar discharge, diagnosis is relatively simple. However, in the majority of cases such discharge is not apparent, so cows need to be examined more closely to identify metritis.

a. Monitoring of Milk Yield

Irrespective of the prime method used for diagnosis of uterine infection, the accuracy of diagnosis can be improved by monitoring milk yield. It is well known that cows with postpartum uterine infections experience a sudden drop in milk yield. However, it is important that a failure of increase in daily milk yield in recently calved cows is an equally important indication of uterine infection (Szenci, 2016). Although monitoring cows in the first 10 days in milk would be sufficient to diagnose most of the cows, a substantial proportion (~20%) would be missed. Hence, a combination of targeted monitoring of all cows at 4, 7, and 12 days in milk in combination with physical examination of cows with milk deviation of more than 12% or failure to increase milk yield at least by 4% (primiparous) or 7% (multiparous) per day in the first 20 days in milk have proven very efficacious in diagnosing cows with metritis and metabolic diseases (ketosis and displaced abomasum) (Galvao, 2011). Although the proposed system may be effective, it requires individual daily milk weights. Others have targeted the first 13 (Benzaquen *et al.* 2007) or 14 days in milk (Galvão *et al.* 2009a) for daily monitoring.

Regardless of the monitoring regimen adopted, compliance with the protocol and skill of the evaluator is paramount to the success of the monitoring program.

b. Rectal Palpation

A morphological examination of the bovine reproductive system provides useful information in the evaluation of infertility in cattle (Gonzales *et al.* 1985) Transrectal palpation is the most common and cheapest technique that is being used for the examination of physiological and pathological conditions of the bovine genital tract (Ahmadi *et al.* 2005). Evaluation of cervix and uterus trans-rectum involves estimation of their size, symmetry, and consistency (Bondurant, 1999; LeBlanc *et al.* 2002a) and in pathological cases of salpinx, transrectal palpation is the most common diagnostic method (LeBlanc *et al.* 2002a). In addition, palpable uterine lumen, dominant palpable ovarian structure including corpus luteum (CL), follicle, or no palpable structures can also be recorded (Sheldon and Dobson, 2004).

The genital tract at 21 to 30 days postpartum is typically classified in one of three categories: normal and with mild or severe endometritis (Bondurant, 1999; LeBlanc *et al.* 2002a). Studer and Morrow (1978) assigned a score to the uterus after a transrectal evaluation; a low rating signified that the uterine horns were 2.5 to 3.75 cm in diameter with a slight thickened feeling; a middle rating signified that the uterine horns were 3.73 to 5 cm in a diameter with a definite thickened feeling. A high score signified a uterine horn diameter of > than 5 cm with grooves and caruncles readily palpable and fluid usually present.

The rectal palpation method has some serious drawbacks and remains a challenge. This method is subjective, insensitive, nonspecific and prone to error as it lacks standardization (Miller *et al.* 1980; De Kruif *et al.* 1982; Gilbert, 1992; Foldi *et al.* 2006; Palmer, 2008), and often fails to account for normal events and variability in uterine involution (Lewis, 1997), or to have any association with reproductive performance (Miller *et al.* 1980; LeBlanc *et al.* 2002a). Also, the uterine size and palpable quality of content may vary between individuals and strongly depends on the stage of the postpartum period. Similarly, another set of studies have observed that infertility cannot be easily detected by a routine clinical examination or by the presence of bacterial or viral isolates (Radoslavov and Savova-Burdarvo, 1982; Gonzales *et al.* 1985) and on certain occasions does not provide sufficient information (Grunert 1979; Stolla and Himmer, 1980; Max *et al.* 1997).

c. Vaginal Examination

Vaginal examination is a practical tool to distinguish healthy from diseased cows with uterine infections. The vaginal examination is a standard part of the breeding soundness and is an essential part of the evaluation of a suitable cow or when ever discharge is noted at the vulvar lips (Roberts, 1986). Vaginal examination is not performed on a routine basis but can be performed to aid in diagnosis if a cow has a fever of unknown origin and no uterine discharge can be produced after palpation of the uterus per rectum.

In cases with vulval discharge, diagnosis is simple. Prior to transrectal examination, the cows are inspected for the presence of discharge. So, the clinical examination of cows begins with the inspection of the vulva, perineum, and tail to observe the character of vaginal secretions. (Leutert *et al.* 2012). In very few cows a fresh discharge can be seen on the vulva, perineum, or tail. In the majority of cases such discharge is not apparent, and the cows therefore need to be examined more closely to identify metritis. Hence, if discharge is not visible externally, cows should be examined through any of the below mentioned manual techniques using all hygienic precautions. The manual examination techniques are considered inexpensive, quick and provide additional sensory information such as detection of vaginal lacerations and detection of odor of mucus in vagina.

Different methods of manual vaginal examination are:

(i) **Rectal palpation:** In this back racking of uterus is done in a gentle manner to force the uterine contents out of the vulva, which are subsequently evaluated.

(ii) **Retrieving of genital secretions:** This is the simplest approach for diagnosis of clinical endometritis involving withdrawal of the mucus using an AI sheath for its inspection (Sheldon *et al.* 2002a).

(iii) **Use of gloved hand:** It is also considered a simple, cheap, and fast non-invasive method. In this a gloved hand or the full length arm is inserted through the vulva and the contents of the vagina are withdrawn manually for examination. Care should be taken to wash the vulva with antiseptic solution (e.g. iodine scrub) and to use a clean well lubricated palpation sleeve (Williams *et al.* 2005).

(iv) **Vaginoscopy:** Vaginoscopy has been considered more sensitive than simple external inspection for detection of purulent discharge (Dohmen *et al.* 1995; LeBlanc *et al.* 2002a). Use of vaginal speculum was the earliest method employed to examine the vagina as an aid in determination of the origin of the exudates (which may stem from the uterus, cervix, or vagina) noted at the vulvar lips (Williams *et al.* 2005). LeBlanc *et al.* (2002b) reported that using vaginoscopy for the detection of pus and cervical contents is the most useful tool for the diagnosis of endometritis. The vaginoscopy also helps to know any physical damage to the vulva in a recently calved cow. Miller *et al.* (1980) concluded that vaginoscopic examination is a much more accurate method for diagnosing clinical endometritis than transrectal palpation.

The vaginoscopic examination to assess the origin of the secretion can be done with a metal speculum with three Polanski type valves or one of plastic, previously disinfected. A sterile metal or transparent barrel with a light source can be equally useful (Runciman *et al.* 2008a; Leutert *et al.* 2012). Other options to use vaginoscopy are by using autoclavable plastic or disposable foil-lined cardboard vaginoscopes, to visualise the mucus flowing out of the cervix, or to use an instrument to withdraw the vaginal contents.

Vaginoscopy correctly predicted the uterine infection in 59 to 82% of cases (Miller *et al.* 1980; Dohmen *et al.* 1995; Williams *et al.* 2005; Leutert *et al.* 2012) but vaginitis, cervicitis, cystitis and purulent nephritis may give false results. Vaginoscopy is convenient to use yet there is risk of disease transmission and the requisite precautions need to be taken into account. The presence or absence of uterine discharge at the time of vaginoscopy may be influenced by the severity of the infection, the uterine contractions, the rate of secretion removal, and by the personnel performing the examination. Leaks in the closed cervix may not be detected; this does not automatically indicate the absence of endometritis (LeBlanc *et al.* 2002a).

An alternative approach for sampling vaginal contents is with the help of the Metricheck device, which consists of a 50 cm long stainless steel rod provided at one end with a 4 cm rubber hemisphere and at the other end with a metal handle (Figure 1).

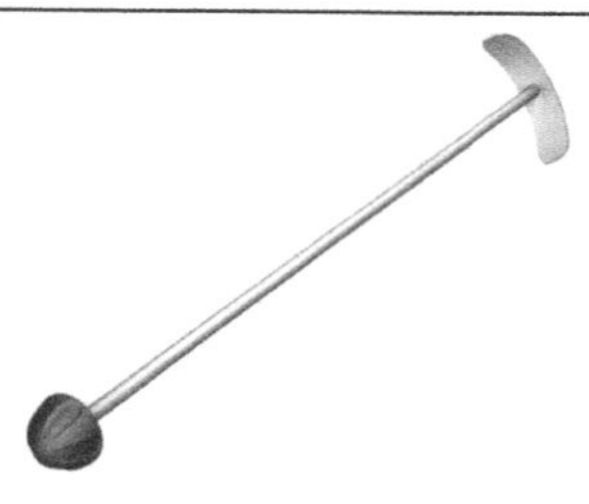

Figure 1: A Metricheck device depicting a rubber hemisphere on a metal handle

The technique consists of introducing the Metricheck device through cleaned vulvar lips, advanced to the cranial extent of the vaginal fornix and then retracted back at an angle of 45 degrees. If no uterine discharges are observed in the hemisphere of the device, the device can be re-introduced until purulent material may be visualized to the convex surface in confirmed.

Amongst the three different methods of vaginal examination viz. vaginoscopy, gloved hand and Metricheck for diagnosing clinical endometritis between Days 21 to 27 after calving, it was observed that somewhat more cows (47.5%) could be diagnosed by Metricheck than by the other two methods (vaginoscopy: 36.9%, gloved hand: 36.8%) (Barlund *et al.* 2008; Pleticha *et al.* 2009).

In the absence of a gold standard it seems that vaginoscopy or Metricheck is preferred to use as a cow-side diagnostic tool for diagnosing clinical endometritis in the field (de Boer *et al.* 2014).

The aforementioned manual examinations done appropriately with all necessary precautions do not cause uterine bacterial contamination, provoke an acute phase protein response, or affect uterine horn diameter (Sheldon *et al.* 2008).

Irrespective of the method employed to collect the genital accumulate, the latter is evaluated to ascertain the presence of uterine infection, its severity, the quantitative and qualitative microbial load and even the prognostic outcome of the treatment as well as the probable time for clinical recovery (Sheldon and Noakes, 1998; Williams and Sheldon, 2003; McDougall *et al.* 2007; Pleticha *et al.* 2009; Peter *et al.* 2011).

Vaginal contents are assessed for color, proportion and volume of pus and odor resulting in uterine infection clinical score. Sheldon and Dobson (2004) confirmed that, the examination of the contents of the vagina for the presence of pus is the most useful method for diagnosis of endometritis. The character (color) and odor of vaginal discharge can be scored and endometritis score is highly correlated with growth density of pathogenic bacteria in the uterus (Mortimer *et al.* 1997).

The character (color) and the odor of the vaginal mucus (Sheldon and Dobson, 2004; Sheldon *et al.* 2006) can be scored as below so as to produce a clinical endometritis score:

Mucus color (Figure 2)

Score 0 = clear or translucent mucus

Score 1 = mucus containing flecks of white or off-white pus

Score 2 = discharge containing ≤ 50% white or off-white mucopu-rulent material

Score 3 = discharge containing ≥ 50% purulent material, usually white or yellow, but occasionally sanguineous

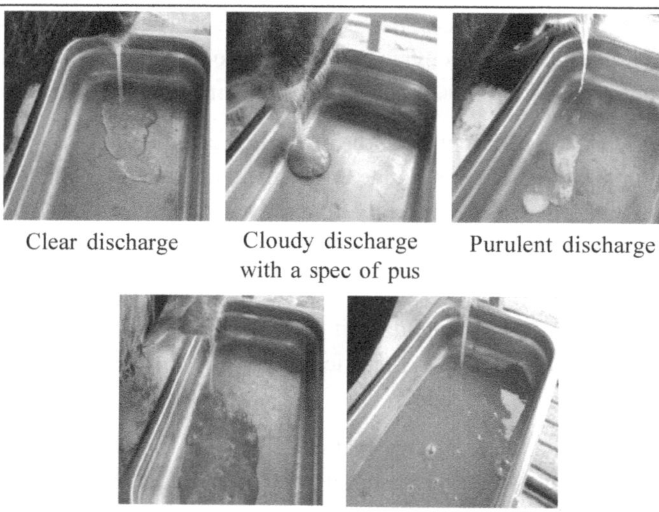

Clear discharge — Cloudy discharge with a spec of pus — Purulent discharge — Bloody and purulent discharge — Reddish brown, watery and fetid discharge

Figure 2: Vaginal discharge scoring system for early postpartum cows

Mucus odor

Score 0 = no unpleasant odor

Score 3 = Fetid odor

The mucus color and odor scores are summed to give an endometritis clinical score ranging from 0 to 6. Although few animals with a mucus color score of <3 also have a fetid odor, weighting the fetid odor score as 3 avoids the potential confusion that might occur if the score was 1. The clinical endometritis score reflects the presence and semi-quantitative load of recognized uterine pathogens, but not other bacteria in the uterine lumen (Williams *et al.* 2005).

The smell reflects the presence of certain bacteria and their growth density in the uterus. Clear mucus denotes a small number of bacteria in the uterus, and purulent discharge with score 3 for a fetid odor was associated with the presence of infection in the uterus. In one study (Williams *et al.* 2005) mucopurulent discharge was associated with *Fusobacterium necrophorum* and purulent discharge was associated with *Arcanobacterium pyogenes* and *Proteus* spp., whilst a fetid odor was associated with a greater load of *Arcanobacterium pyogenes*, *E.coli*, *Streptococci*, and *Mannheimia haemolytica*. Using a semi-quantitative method, the median bacterial growth density of pathogenic bacteria in the uterine lumen was higher for animals with purulent (4 *vs.* 1, $P < 0.05$) or fetid (4 *vs.* 1, $P < 0.05$) material in the vagina than normal animals. In addition, endometritis clinical score is prognostic for the likely success of treatment (Sheldon and Noakes, 1998; LeBlanc *et al.* 2002a; Abdel *et al.* 2016). The success rate for cure of endometritis over a 2 week period, as determined by achieving a final vaginal mucus score of 0, was 44% if the vaginal mucus was purulent with a fetid odour at the start of treatment, but 78% if there were only flecks of pus in the mucus.

d. Microbial Presence

There are microbial communities in and / or on the bodies of all multicellular organisms, which have a significant impact on the biology of its host (Appiah *et al.* 2020). The genital tract of the cows is also not spared and is known to harbor a variety of microbes. Microbial infection may cause the inflammation of the endometrium and may be associated with denudation of the uterine mucosa and alteration of the secretion with early embryonic death, thereby affecting fertility (LeBlanc *et al.* 2002a). Uterine microbial pattern dictates the (i) appearance of genital discharge (Williams *et al.* 2005) (ii) severity of infection (Pascottini *et al.* 2020) (iii) therapeutic response (Jeon *et al.* 2018) and (iv) elicits a change in spectrum from calving till the resumption of normal reproduction rhythm / estrous cycles (Williams *et al.* 2007). For instance, *Bacillus* spp. is typically found in decaying tissue like the lochia of a uterus during involution (Wise, 2016).

The presence of microbes becomes inevitable after parturition and may persist to cause to uterine infections in certain percentage of cows (Borsberry and Dobson, 1989; Sheldon *et al.* 2002b). In particular, bacterial species such as *Trueperella pyogenes* and *Escherichia coli* have been reported to be pathogenic to the uterus (Williams *et al.* 2005). However, a clear relation between the detection rates of bacteria and severity of uterine inflammation has not been established. Occurrence of uterine infection in the dairy cows is neither casual nor indirect, but multifactorial, involving interaction between the host and pathogens, which may vary between different individuals (Adnane and Chapwanya, 2022). The variation can be attributed to differences in the environment, host niches and health status of the cows (Appiah *et al.* 2020; Adnane and Chapwanya, 2022). It is known since long that the environmental pathogens entering into vagina have the ability to move through the cervix into the bovine uterus (Sheldon, 2004). However, several recent studies propose an "endogenous route hypothesis", whereby the microbial entry or colonization in the uterus may also occur from gut through hematogenous route (Galvao *et al.* 2009a; Appiah *et al.* 2020). Hence, apart from cyclic cows, the pregnant uterus in the cows may also not be microbiologically sterile. Using a deep sequencing metagenomic approach, *Fusobacterium necrophorum*, *Porphyromonasslevii* and *Trueperella pyogenes* have been detected in pregnant uterus in cows (Karstrup *et al.* 2017).

Microbial culture-based studies have isolated nearly 200 different kinds of microbes in cows exhibiting reproductive infection (Drillich and Wagener, 2018). Not all of these bacteria, however, are considered as pathogens. The role of bacteria that can be regarded as opportunistic or potential pathogens, e.g. *Bacillus pumilus*, is still under investigation.

In terms of microbiological variety, *Escherichia coli*, *Trueperella pyogenes*, *Streptococcus* spp., *Pasteurella* spp., *Prevotella* spp. and *Fusobacterium necrophorum* were the most common isolates detected from the uterine lumen of post-partum cows (Elliot *et al.* 1968; Griffin *et al.* 1974; Studer and Morrow, 1978; Bonnett *et al.* 1991a; Dohmen *et al.* 1995; Janeway Jr *et al.* 2001; Sheldon *et al.* 2002b; Sheldon, 2004; Williams *et al.* 2005). In other studies on confirmed cases of uterine infection postpartum, *Escherichia coli* and *Trueperella pyogenes*, followed by anaerobic bacteria such as *Prevotella* spp., *Fusobacterium necrophorum*, and *Fusobacterium nucleatum* were the most frequent microbiological isolates (Dohmen *et al.* 2000; Gilbert *et al.* 2005; Williams *et al.* 2005; Sheldon *et al.* 2006; Prunner *et al.* 2014a,b; Madoz *et al.* 2014). The postpartum uterus favors an anaerobic environment (Elliot *et al.* 1968; Griffin *et al.* 1974; Sheldon *et al.* 2002b; Clemmons *et al.* 2017), as against aerobic environment, at least during estrus, in cycling cows (Dhaliwal *et al.* 2001). The difference in uterine environment can be the reason for finding more aerobic microbes in the genital discharge of estrus than the postpartum cows (Ballas *et al.* 2021). Accordingly, considering the microbial spectrum in the endometritis cycling cows, a comparison after reviewing 7 different studies indicated aerobes or facultative anaerobes such as *Escherichia coli* followed by *Staphylococcus* spp., *Streptococcus* spp., *Bacillus* spp., *Corynebacterium* spp., *Pseudomonas* spp. to have dominated their uterine presence to a variable extent in repeat breeding or endometritis affected cycling cows. The respective percentages for the aforesaid microbes ranged from 31 to 40%, 27 to 33%, 9 to 15%, 2 to 15%, 2 to 18% and 4 to 17%, respectively (Singla *et al.* 2004; Behera *et al.* 2015; Desai *et al.* 2017; Manjhi *et al.* 2019; Painuly, 2019; Pascottini *et al.* 2020; Sikrodia *et al.* 2021). In addition, *Micrococcus* spp. has also been detected in uterine infections postpartum and in cycling or repeat breeding cows (Panangala *et al.* 1978; Williams *et al.* 2005; Sharma *et al.* 2008; Raval *et al.* 2018) whereas, *Flavobacterium* spp. was detected in cows affected with uterine diseases (Rodrigues *et al.* 2015).

There is evidence that a specific strain of *E. coli*, EnPEC, is the main pathogen for metritis (Sheldon *et al.* 2009b). On the basis of aerobic and anaerobic uterine cultures Sheldon *et al.* (2002b) and Williams *et al.* (2007) categorized the microbes as recognized uterine pathogens, potential uterine pathogens and opportunistic uterine contaminants, respectively. *Trueperella pyogenes, Prevotella melaninogenica, Escherichia coli, Fusobaterium necrophorum and Proteus* spp. were considered as the recognized uterine pathogens that cause endometrial lesions. On the other hand, *Bacillus licheniformis, Enterococcus faecalis, Mannhiemia haemolytica, Pasteurella multocida, Peptostreptococcus* spp., *Staphylococcus aureus and n*on-haemolytic *Streptococci* were categorized as potential uterine pathogens not commonly associated with uterine lesions. *Clostridium perfringens, Klebsiella pneumoniae, Micrococcus* spp., *Providencia stuartii, Coagulase negative Staphylococcus* spp., *alpha-Haemoltyic Streptococci, Streptococcus acidominimus,* and *Aspergillus* spp. were considered as opportunistic uterine contaminants that were routinely isolated from the bovine uterus and were not associated with endometritis.

The physical appearance of the genital discharge varies with the severity of uterine infections. The cows exhibiting presence of recognized and potential category of uterine microbes, as listed *vide supra*, have revealed presence of mucopurulent or purulent and fetid vaginal discharge in 21 to 28 days postpartum cows (Williams *et al.* 2005; Wagener *et al.* 2015; Pascottini *et al.* 2020). To the contrary, a discord between uterine damage and purulent discharge has also been projected. Uterine histology was found to be normal in 40% cows exuding purulent uterine discharge (Hartigan *et al.* 1972).

It would be clinically relevant to know the association of uterine microbiota as per their expected pathogenic potential in causing uterine disease of different severities, which in descending order would be metritis, clinical endometritis and subclinical endometritis. In other words anticipating a relationship between severities of uterine infection with mere presence of bacterial communities would be inappropriate. For instance, *Escherichia coli* virulence factors, such as fimH, hlyA, cdt, kpsMII, ibeA, and astA, are associated with uterine diseases and impaired reproductive performance (Bicalho *et al.* 2010). It is noteworthy that the bacterial infection of uterine lumen during first two weeks after parturition is ubiquitous. Around 93% samples were positive for bacterial culture after parturition which reduced to 9% by 7 to 8 weeks after parturition (Elliot *et al.* 1968). The role of pathogenic and potential bacteria in subclinical endometritis is unclear. No association was found between the bacterial culture results and subclinical endometritis at the time of diagnosis (Pascottini *et al.* 2020). Hence, a pathogenic effect by the opportunistic uterine microbes cannot be precluded (van der Burgt *et al.* 2007; Nicholas *et al.* 2008; Adnane and Chapwanya, 2022). Another set of studies have showed that the bacterial populations from the uterus of cows with subclinical endometritis were similar to those of healthy cows, but differed from the cows with clinical endometritis (Lazzari *et al.* 2011; Madoz *et al.* 2014; Marino *et al.* 2017; Pascottini *et al.* 2020). In support to the latter studies, neither *Trueperella pyogenes* nor *Escherichia coli* that were associated with metritis and clinical endometritis could be detected in the pathogenesis of subclinical endometritis (Prunner *et al.* 2014a,b; Gartner *et al.* 2015; Ballas *et al.* 2021).

(i) Dynamism of uterine microbiota: The uterine microbiota is a dynamic consortium, particularly in fresh postpartum cows which, however, gets stabilized by the fifth week of calving (Santos and Bicalho, 2012). Nonetheless, the core bacterial community remained stable in healthy, clinical endometritis and subclinical endometritis cows (Santos and Bicalho, 2012; Pascottini *et al.* 2021). Desai *et al.* (2017) also reported an inter-cyclic variation in microbial spectrum in 17% of the cows with subclinical endometritis. Several other studies have also reported undergoing brisk changes by microbial communities in the genital tract (Otero *et al.* 1999; Ault *et al.* 2019a,b; Quereda *et al.* 2020; Gohil *et al.* 2022). Hence, prior to fifth week postpartum there are chances of contamination and recontamination.

(ii) Single *versus* mixed bacterial isolates: A direct connection of single *versus* mixed genital infection with severity of uterine disease has not been established. The available literature, however, is suggestive of presence of mixed infection in cows with severe uterine disease. Mixed infections are known to harbor postpartum metritic cows (Elliot *et al.* 1968; Griffin *et al.* 1974; Sheldon *et al.* 2002b; Williams *et al.* 2005; Galvao *et al.* 2009b). Moreover, averaged for different studies, single bacterial infection was recorded in nearly 64% cows with subclinical endometritis as compared to 54% in cows with clinical endometritis (Abd El-Kader and Sehata, 2001; Sood and Ahuja, 2023; Sharma *et al.* 2008; Behera *et al.* 2015; Crivei *et al.* 2016; Manjhi *et al.* 2019).

(iii) Relation between microbes and uterine health: *Escherichia coli*, *Trueperella pyogenes*, *Fusobacterium necrophorum*, and *Prevotella melaninogenica* are thought to be the primary pathogens causing uterine disease, because these bacteria were frequently isolated from the uterine lumen of diseased cows, and they were shown to affect endometrium and ovary function (Griffin *et al.* 1974; Dohmen *et al.* 1995; Sheldon *et al.* 2002b; Miller *et al.* 2007 ; Williams *et al.* 2007; Bicalho *et al.* 2012). In particular, *E. coli* emerged in the uterus within a few days of calving, followed by the colonization of *T. pyogenes and F. necrophorum*, suggesting that the uterine pathogens work synergistically (Williams *et al.* 2005). *T. pyogenes*,

F. necrophorum, and Bacteroidetes, particularly *Porphyromonas levii*, but also *Prev. melaninogenica*, were correlated among themselves, and were isolated with higher frequency as the uterine discharge score worsened (i.e., from flecks of pus to hemorrhagic and fetid) (Dohmen *et al.* 1995).

A. pyogenes produces a growth factor for *F. necrophorum*, *F. necrophorum* produces a leukotoxin, and *P. melaninogenicus* produces a substance that inhibits phagocytosis (Sheldon and Dobson, 2004). In addition, proceeding of uterine infection with *E. coli* has also been followed possibly with bovine herpesvirus 4, which has been linked with clinical disease severity and impact on fertility (Dohmen *et al.* 2000, Williams *et al.* 2007, Donofrio *et al.* 2009). Bovine herpesvirus is the only virus commonly associated with uterine infections that develops in the uterus after gestation, being widespread in the European Union (Castrucci *et al.* 1986).

In regards to bacteria associated with uterine health, Bonnett *et al.* (1991a) reported that isolation of α-hemolytic streptococci was negatively correlated with isolation of *A. pyogenes* and with inflammation. Dohmen *et al.* (1995) reported that streptococci were isolated less frequently as the uterine discharge worsened, and that streptococci were negatively associated with the presence of *T. pyogenes* at 14 days postpartum. Williams *et al.* (2005) observed that cows with α-hemolytic streptococci, the most common opportunist, and coagulase-negative staphylococci were less likely to be diagnosed with endometritis. Huszenicza *et al.* (1999) also reported a lower prevalence of streptococci and staphylococci in cows with metritis compared with healthy cows. Gilbert and Santos (2016) reported that cows with α-hemolytic streptococci at 7 days postpartum had improved fertility. Therefore, culture-based studies indicate that streptococci, or more specifically, α-hemolytic streptococci, can promote uterine health. There were also some indications that *E. coli* detected later in lactation could promote uterine health (Bonnett *et al.* 1993; Dohmen *et al.* 1995), but these findings were inconsistent (de Boer *et al.* 2015b).

(iv) Relationship of microbes at estrus with ensuing pregnancy

(a) Uterine microbiota after treatment: Using lipopolysaccharide, with or without autologous plasma, to treat subclinical endometritis, the post-treatment genital discharge became bacteriologically sterile in 70 to 80% of the cows (Singla *et al.* 2004; Desai *et al.* 2017). Similarly, antibiotic usage although reduced the microbial load, but failed to eliminate the bacterial population fully in metritis cows (Jeon and Galvao, 2018). Also, antimicrobial therapy cannot sterilize the uterus nor prevent the continual recontamination of uterus. Successful resolution of uterine infections requires to be complemented by an effective uterine defense for clearing of infection, which possibly decides the therapeutic outcome or success (Paisley *et al.* 1986).

(b) **Pregnancy in presence of bacteria:** It has been hypothesized that the bacterial load or spectrum absolutely does not dictate the uterine health status in cows in terms of pregnancy outcome (Bicalho *et al.* 2016). In consonance, association between specific bacteriological findings and subsequent insemination success revealed a conception rate of 25.8% in cows with mild clinical endometritis (presence of turbid or pus flecks in genital mucus), compared to 43.1% in reproductively healthy cows (Ballas *et al.* 2021). Hence, presence of bacterial load in mild clinical endometritis although had a negative effect on the likelihood of pregnancy, but conception failure in all such cows was not a thumb rule. In the same study, Ballas *et al.* (2021), found that among the 49 different genera isolated in both healthy and mild clinical endometritis cows, *Bacillus* spp. and *Corynebacterium* spp. were significantly less in cows with subclinical endometritis. Hence, it is plausible that in addition to microbiota composition, several cow linked intrinsic factors may play an important role in deciding the ability of the cows to get pregnant in presence of uterine microbes (Santos and Bicalho, 2012; Wagener *et al.* 2015; Bicalho *et al.* 2016).

(c) **Gram positive *versus* gram negative bacterial isolation:** The available literature suggests nearly equal distribution of gram positive and gram negative bacteria both in clinical endometritis (Sood *et al.* 2002c; Zahid, 2004; Crivei *et al.* 2016) and subclinical endometritis cows (Brodzki *et al.* 2015; Painuly, 2019).The gram staining characteristics of uterine microbes are relevant in terms of severity of disease (Kaneko and Kawakami, 2008; Sheldon *et al.* 2010; Bicalho *et al.* 2016) as well as the effect on hypothalamus-pituitary-gonadal axis (Dahiya *et al.* 2018; Sheldon *et al.* 2019). In comparison to gram positive, gram negative bacteria are more deleterious due to a distinctive cell wall configuration (Moyes *et al.* 2009) and are more resistant (Breijyeh *et al.* 2020). The gram positive and gram negative bacteria also differ in executing an undesirable carryover effect on reproduction. For instance, gram negative bacteria, such as *Escherichia coli,* triggers an immediate and more severe effect, whereas gram positive bacteria, such as *Staphylococcus aureus,* has both immediate and long-term effect in disrupting structural and steroidogenic function of ovarian follicles (Lavon *et al.* 2011).

(d) **Microbial spectrum in relation to severity of uterine disease:** Several studies have shown that bacterial populations isolated from the uterus of cows diagnosed with subclinical endometritis was nearly similar to those of healthy cows which, however, differed from those with clinical endometritis (Lazzari *et al.* 2011; Madoz *et al.* 2014; Marino *et al.* 2017). Presence of *Trueperella pyogenes* in the fresh postpartum uterus of cows was a risk factor for development of clinical endometritis, but neither *Trueperella pyogenes* nor *Escherichia coli* were associated with subclinical endometritis (Prunner *et al.* 2014a,b). Pascottini *et al.* (2020) found that at the genus level, cows with clinical endometritis had microbiota characterized by greater relative abundance of *Fusobacterium* and *Trueperella*, lower relative abundance of *Escherichia, Shigella, Lactobacillus, Prevotella, Schlegelella, Staphylococcus,* and *Streptococcus* than the healthy and subclinical endometritis cows. In other words, the common pathogens associated with metritis and clinical endometritis do not have a significant role in the subclinical endometritis pathogenesis. Commensal bacteria, instead, could be of vital importance in subclinical endometritis (Gartner *et al.* 2015). The discrepancy among different studies on the microbial involvement in subclinical endometritis is justified from the research conducted to date proving that the composition of uterine microflora in a community varies widely between environmental sites, host niches, and health status of the cows (Sheldon *et al.* 2006; Herath *et al.* 2009; Carneiro *et al.* 2016; Sheldon *et al.* 2019, 2020). To the contrary, subclinical endometritis without any evidence of infection is also on record (Lazzari *et al.* 2011; Marino *et al.* 2017).

(v) **Techniques used for microbial detection**

(a) **Bacterial culture:** A uterine culture is an essential tool to determine the etiology of uterine infection (Studer and Morrow, 1978; Olson *et al.* 1986; Bretzlaff, 1987; Lewis, 1997; Bondurant, 1999; LeBlanc *et al.* 2002a; Sheldon *et al.* 2004 a). A swab is the most accurate means of obtaining samples for identification of the specific and non-specific bacteria that cause infection (Olson *et al.* 1986; Noakes *et al.* 1991; Bonnett *et al.*1993). A true uterine culture should be taken from the uterus without contamination by extraneous bacteria. Yet the difficulty in obtaining samples representative of the uterus limit the frequent use of such approach. Extra corporated vulvar and vaginal contaminations are all potential sources of bacteria that can add to the vagueness in the interpretation of the results (Dohmen *et al.* 2000; Sheldon *et al.* 2004 a). For instance, the swabs collected from uterine lumen of dairy cows 21 and 28 days after calving, purulent vaginal mucus was associated with growth density of pathogenic bacteria, but not the opportunist contaminants, which includes *Arcanobacterium pyogenes*, *Fusobacterium necrophorum*, *Prevotella* spp. and *Bacteroides* (Dohmen *et al.* 1995; Williams *et al.* 2005; Westermann *et al.* 2010).

It is not surprising to find no bacterial isolates using traditional culture based bacteriological isolations in the cows with different grades of uterine infections. Certain percentage of cows with clinical endometritis (Ahmadi *et al.* 2015; Pascottini *et al.* 2020) and especially those with subclinical endometritis (Behera *et al.* 2015; Manjhi *et al.* 2019; Pascottini *et al.* 2020) have been found to be bacteriolocially negative. After cytological confirmation, 18%, 28% and 20% bacterial samples were found negative in healthy, subclinical endometritis and clinical endometritis cows, respectively (Pascottini *et al.* 2020). In another study, 47% samples were found to be negative for bacteria in cows with clinical endometritis (Ahmadi *et al.* 2015). Similarly, 6.9% and 9.7% uterine samples in the healthy and subclinical endometritis cows were bacteriologically negative (Ballas *et al.* 2021). Madoz *et al.* (2014) found no growth in all subclinical endometritis cows as against presence of growth in 37.5% cows with clinical endometritis. All the three mentioned studies utilized aerobic and anaerobic culture conditions in ascertaining the microbial presence. Few other Indian studies, exclusively relying on aerobic cultures, have reported up to 60% of the genital discharge samples as negative for bacterial growth in subclinical endometritis cows (Behera *et al.* 2015; Yadav *et al.* 2018; Sikrodia *et al.* 2021). In another study, 100% normal cows, as against 20% cows with subclinical endometritis revealed absence of bacterial growth in the genital discharge (Manjhi *et al.* 2019). Lack of anaerobic bacterial conformation can be the reason for relatively more number of cows with no growth in the genital discharge samples in the latter set of studies. Apart from the bacterial isolation approaches, the other plausible explanations of finding no bacterial growth in uterine infections would be that by the time clinical endometritis and subclinical endometritis are diagnosed, the host defense mechanism, still operational, removes the bacteria from the uterus and restores a normal uterine environment. Hiding of bacteria behind inconspicuous cell morphology (biofilm) and co-localization in the lamina propria of the uterus, could be the other reasons (Karstrup *et al.* 2017; Raheel *et al.* 2020). Also needs to be taken into the account is that the subclinical form of endometritis may persist after the treatment of clinical endometritis, as also a deranged immune competence, or tissue repair (Sheldon *et al.* 2019), which may reveal absence of bacterial commodities from the uterus.

It also needs to be understood that bacteriological findings provide only evidence for infection, but not for inflammation.

(b) Metagenomics: Under standard laboratory conditions, less than 1% of the microorganisms in many environments are readily cultured (Amann *et al.* 1995; Whitman *et al.* 1998). Bacterial diversity in healthy and metritic uteri was greater and more complex as observed using metagenomic analysis than described previously for traditional culture methods (Santos *et al.* 2011). Among different reasons, as mentioned previously, for finding a negative microbial growth in cows suspected for uterine infection, inability to detect microbes using standard microbial culture practices appears to be the most prominent reason. Also, there is distinct evidence that the uterine microbes in some cows with low grade uterine infection remained elusive in the genital discharge after routine culture methods (Madoz *et al.* 2013). In the past few decades, the use of advanced metagenomics has led to an explosion in understanding the bacterial microflora compositions in various segments of reproductive tract (Table 1). In addition, the recent rebirth of culture techniques, in addition to bacteria that have so far only been detected using molecular tools, has enabled the isolation of fastidious and new types of bacterial species in the vagina, cervix, and uterus. Comparing bacteria in the reproductive microbiome using advanced molecular tools with those isolated by traditional culture-based methods has shown that these two different techniques enabled an extensive, but somewhat overlapping range of vaginal, uterine, and cervical bacteria species. Dysbiosis or infections caused by unsteadiness in the reproductive tract microflora using metagenomic analysis can provide more appropriate treatments. The first study on uterine metagenomics in cows comprising of healthy *versus* metritic subjects within 21 days after calving was undertaken by Santos *et al.* (2011). They found a great variation in bacterial diversity and were able to identify certain microbes not isolated earlier by traditional culture methods. A subsequent metagenomic study confirmed a dynamic change in bacterial spectrum over a period of time after calving (Peng *et al.* 2013). Metagenomic analysis of uterine microbial biodiversity enabled to find a reduction in certain groups of bacteria 35 days after calving in cows' supplemented mineral mixture in advanced stages of gestation (Machado *et al.* 2012b).

In metagenomic analysis, *F. necrophorum* was found to be the most abundant species in the uterus, while *P. melaninogenica*, *E. coli*, and *T. pyogenes* were found to be extremely rare (<1% abundance) with no association with metritis (Jeon *et al.* 2015, 2016). Conversely, *E. coli* was shown to be associated with uterine health, which conflicts with results from previous studies based on culture methods (Jeon *et al.* 2015).

Gram-positive facultative anaerobic *Helcococcus ovis* and *Helcococcus kunzii* were isolated from the uterus of dairy cows with metritis (Bilk *et al.* 2011; Locatelli *et al.* 2013); but their involvement in metritis was not evident. Later metagenomic studies found that *Helcococcus* was associated with metritis (Jeon *et al.* 2015) and endometritis (Machado *et al.* 2012b), suggesting it to be a potential uterine pathogen. This is also supported in a study by Cunha *et al.* (2018) in which *H. ovis* was significantly prevalent in metritic than in the healthy cows. Because of its presence together with *E. coli* and *T. pyogenes* (Locatelli *et al.* 2013), interaction with other uterine pathogens is likely during metritis development.

Table 1: Metagenomic analysis of uterine microbiota to identify pathogenic bacteria associated with metritis in postpartum dairy cows

Study	**Bacteria (taxonomic categories)**
Santos *et al.* (2011)	*Fusobacteria* (phylum)
Machado *et al.* (2012b)	*Ureaplasma* (genus), *Bacteroides* (genus)
Peng *et al.* (2013)	*Bacteroidetes* (phylum)*Peptostreptococcus* (genus) *Fusobacterium* (genus
Jeon *et al.* (2015, 2016, 2018)	*Bacteroides* (genus)*Porphyromonas* (genus)*Fusobacterium* (genus)
Knudsen *et al.* (2016)	*Porphyromonadaceae* (family)*Fusobacteriaceae* (family)
Bicalho *et al.* (2017)	*Bacteroidetes* (phylum)*Fusobacteria* (phylum)
Sicsic *et al.* (2018)	*Bacteroides* (genus)*Porphyromonas* (genus)*Fusobacterium* (genus)

In another study, metagenomics also revealed similar groups of bacteria in the Nellore, Holstein, and Fleckvieh cattle. Such bacteria at the phylum level specifically comprised *Firmicutes*, *Bacteroidetes*, and *Proteobacteria* (LeBlanc *et al.* 2002a; Nightingale *et al.* 2015), and that constitute about 30 to 40% of the total bacteria present in the vaginal niche (Opsomer *et al.* 2000). Another study by Giannattasio-Ferraz *et al.* (2019) also revealed a substantial similarity existing between the vaginal microbiota of Nellore cattle and Gyr breed, and the main bacterial species described were mainly *Aeribacillus*, *Bacillus*, *Clostridium*, *Bacteroides*, and *Ruminococcus*. *Mycosphaerella* was the dominant fungus in this microbiota, while *Archaea* were poorly represented, and *Methanobrevibacter* were the most common. These data start to unravel the bovine vaginal microbiota, and a broad picture is being defined with the primary microbial agents from reproductive tract, which permit studies of probiotics and prebiotics.

(c) **Fourier- transform infrared spectroscopy (FTIR):** FTIR spectrometers rely on the same basic principle as non-dispersive infrared (NDIR) analyzers, i.e., the fact that many gases absorb infra red radiation at species-specific frequencies. However, FTIR spectroscopy is a disperse method, which is suggestive that the measurements are performed over a broad spectrum instead of a narrow band of frequencies. Due to uncertainty of diagnosing uterine microbes using conventional microbiological techniques, the understanding of the uterine microbiota and its interactions is increasing with the use of modern high-resolution techniques such as FTIR. FTIR, a vibrational spectroscopy method that provides information about the biochemical composition (proteins, lipids, nucleic acids, polysaccharides, etc.) of the analyzed materials (Naumann *et al.* 1991; Ehling-Schulz *et al.* 2005), and has been established as a powerful method for bacterial identification (Kuhm *et al.* 2009; Jaureguiberry *et al.* 2016). As microorganisms differ biochemically, the spectra provided by this method serve as a fingerprint, allowing differentiating between genus, species, strains and serotypes (Oberreuter *et al.* 2002; Bosch *et al.* 2010; Grunert *et al.* 2013). Spectral distances are indicative of the biochemical difference between microorganisms. This information is used in databases as references for future analyses of isolates. Thus, databases containing reference spectra of different strains play a key role in bacterial identification by FTIR spectroscopy. With an appropriate reference database, FTIR spectroscopy is a simple, relatively fast and inexpensive method for the identification of bacteria. Although FTIR spectroscopy is not an on-farm test, which would be preferred for field practice, time from sampling to diagnosis was approximately one day shorter than for routine methods of bacteriological identification. Furthermore, FTIR spectroscopy is a high throughput technique, i.e. it takes less work per sample (Fernandez Olmos *et al.* 2010).

e. Endometrial Biopsy

Endometrial biopsy is an important key to identify the nature of an infertility problem (Bretzlaff, 1987). When combined with the results of microbiology, it permits a more accurate assessment of cows' reproduction prognosis (Studer and Morrow, 1978; Javed and Khan, 1991). In addition to microbial culture, use of biopsy in conjunction with a detailed history, rectal and vaginal examinations can lead to a more accurate prognosis of infertility and greater therapeutic efficiency (Kasimanickam *et al.* 2004; Barlund *et al.* 2008; Oral *et al.* 2009). Uterine biopsy allows for endometrial morphologic analysis and may correlate with bovine infertility, assuming the sample is representative of both uterine horns (Bonnett *et al.* 1991c; Rodriguez *et al.* 1991).

Uterine biopsy is commonly used in human medicine as it is considered the gold standard for evaluating the human endometrium (Strowitzki *et al.* 2006). In domestic animals, uterine biopsy has been used since the 1960s to investigate causes of infertility in mares (Brandt and Manning, 1996), and it is a routine diagnosis method today (Nielsen, 2005). However, in dairy cattle uterine biopsy is rarely performed by practitioners, and it is almost exclusively used for research purposes. The limited use of biopsy in clinical practice may be related with inconveniences associated to sampling, time taken for processing, requirement for laboratory skills, laboratory costs, and time to report (Nielsen, 2005) and also to the risk of inducing endometritis and the subsequent detrimental effect on subsequent fertility (Kasimanickam *et al.* 2004; Miller *et al.* 1980; Bonnett *et al.* 1993). While previously, de Bois and Manspeaker (1986) and Chapwanya *et al.* (2010) recommended the use of endometrial biopsy as a safe procedure causing no adverse effects on cow's reproductive capacity, and the haemorrhages are of little or no clinical significance so that the resulting lesions from the biopsy heal rapidly. Others consider uterine biopsy to be an invasive technique, and the possibility exists of impairment in reproductive performance caused by biopsy. Bonnett *et al.* (1993) showed that it might impair reproductive performance. The latter study used a punch that collected a larger fragment than the one collected using the Yeoman biopsy forceps used by De Bois and Manspeaker (1986) and Chapwanya *et al.* (2010). Uterine biopsy may cause an increase in days open; this work proved that it is a very specific diagnostic method (Meira Jr. *et al.* 2012). Also, there is a lack of agreement between biopsy and cytology findings. Few studies using uterine biopsy for the diagnosis of subclinical endometritis in dairy cattle (de Boer *et al.* 2014), found a poor agreement of biopsy with the cytology findings (Madoz *et al.* 2014; Marino *et al.* 2017, Meira Jr. *et al.* 2012). The histopathological examination of biopsy samples gives detailed information about the degree of inflammation, distribution of the inflammatory infiltrate, or the lesions that may exist, whereas cytology only assesses the superficial layer of the endometrium (Meira Jr. *et al.* 2012). Thus, it is not surprising that direct comparison between biopsy and cytology results showed low agreement. Evaluation criteria for diagnosis of subclinical endometritis on biopsy samples, as for cytology, should be established based on detrimental effect on subsequent reproductive performance rather than on the presence of an arbitrary number of inflammatory cells (Meira Jr. *et al.* 2012)

Histologically, endometritis is characterized by disruption of the surface epithelium, infiltration with inflammatory cells, vascular congestion, and stromal edema, with varying degrees of lymphocyte and plasma cell accumulation in the superficial layers (Kennedy and Miller, 1993; BonDurant, 1999).

Procedure: An endometrial tissue is readily obtained with a biopsy punch. Haupner device, a tool most commonly used to collect biopsy in cattle, is guided in the cervix by transrectal manipulation (Chapwanya *et al.* 2009). Albuchin's uterine biopsy catheter is another device used for the purpose. Ideal sample size is at least a 10 to 20 mm x 3 mm to ascertain a complete detail of both the epithelial cell layer and the granular architecture (de Bois and Manspeaker, 1986). Proper disinfection and sterilization of the biopsy instrument are necessary to prevent microbial contamination. Before taking biopsy, thoroughly scrub and clean the vulva and perineal area. Evert the vulval lips and introduce the biopsy instrument in closed position through vagina into uterus. Gently push the piston to open the cutting edge. Press a portion of the uterine wall into the cavity of the cutting edge. Pull the piston caudally to close the cutting edge so as to remove a piece of the endometrium. Withdraw the instrument out of the reproductive tract in closed position. Remove the endometrial tissue from the instrument and immediately transfer it into 10 % neutral buffered formalin solution at room temperature. Tissues are trimmed, dehydrated, cleared and embedded in paraffin sections and cut at a thickness of 5 to 6 micron and stained with haematoxilin and eosin stain for histological examination. Specimens should be taken from both the uterine horns and body due to variability of pathology in each section.

The histological findings are based on the assessment of epithelium height (columnar, cuboidal or flat), epithelial damage (mild to ulceration), inflammatory cell type (mononuclear, PMN), infiltrate intensity (mild to severe), characteristics of lamina propria (inflammatory cell type, infiltrate intensity, lymphocyte aggregates), endometrial gland (atrophied or fibrosed) and vascular changes (vessel degeneration, hemorrhage, hemosiderin macrophages). Factors have been defined that provide a prognosis for cows' fertility. These factors include inflammation, periglandular fibrosis, and cystic glandular degeneration (Singh *et al.* 1983). Endometrial inflammation is evidenced by increased numbers of inflammatory cells in foci or diffused in various areas of the lamina propria. The cellular infiltration is classified by the predominant cell type. Acute inflammation contains mostly PMN cells in blood vessels; their presence during physiologic estrogenic phase of the cycle needs to be taken into account (Bonnett *et al.* 1991b). Neutrophils may be present in high numbers during normal estrum erroneously suggesting acute endometritis. On the other hand their presence during the luteal phase is definitely indicative of an acute endometritis. The initial phases of endometritis is characterized by diffused and possibly the periglandular and perivascular infiltrations that are dominated by neutrophils and lymphocytes. Most commonly seen in chronic endometritis is infiltration of lymphocytes, accompanied by various degrees of plasma cell and macrophage infiltration (Kubar and Jalakas, 2002). Endometrial fibrosis occurs mostly around the gland and appears in response to inflammation of glandular damage (Dzhuroua and Gulubinov, 1981; Bollo *et al.* 1990). Periglandular fibrosis may interfere with gland function to the extent that glandular support of the early conceptus is altered and cause early embryonic death (Roberts, 1986; Noakes *et al.* 2002). Moderate and severe cases of endometritis are much easier to diagnose on the basis of the increased number of inflammatory cells spread throughout the stratum compactum and spongiosum layers compared with few cells seen in mild endometritis.

Overall, 60 to 70% of the biopsies may be readable (Bonnett *et al.* 1991a,b,c; Meira Jr. *et al.* 2012) Bonnett *et al.* (1993) found sensitivity of 92% and specificity of 77% for biopsy, using pregnancy at 120 d postpartum as a reference. Another study found the sensitivity of 44% and specificity of 92% for the diagnosis of subclinical endometritis using cytology as a gold standard (Meira Jr. *et al.* 2012). The interpretation of uterine biopsy may also depend on the extent of involvement of endometrial glands in the inflammation process during subclinical endometritis. In the said case, excluding the consideration of glandular inflammation, as revealed by absence of abnormal genital discharge, increased the diagnostic efficiency to 78% and specificity to 94%, respectively. This suggests a need for identifying the type of glandular lesions having real importance for subclinical endometritis.

Almeida *et al.* (1986) found no differences in epithelia obtained from either caruncular or intercaruncular endometrium, only that gland openings were absent in the caruncular tissue. However, results in their study and elsewhere suggest that the bovine intercaruncular endometrium is a far more physiologically active structure. The uterine biopsy in spite of having been considered as a gold standard by some does not seem to be having perfect sensitivity and specificity. The histological inferences can be more reliable in mares or primates with a uniform endometrial structure throughout the uterus; not so in ruminants, where there are measurable differences between caruncular and intercaruncular endometrium. A potential confounding factor is the fact that the basal zone of the endometrial glands in the bovine encroaches at the base of the aglandular caruncle. Whilst this is no problem when describing the bovine endometrium generally or the intercaruncular regions specifically, those glands adjacent to the caruncle must be described when using the stereological method. Using stereology, the differences in endometrial anatomy between site of sampling and either follicular or luteal phases of the estrous cycle were always more significant in heifers than cows. The endometrial thickness in cows was always greater than for heifers, irrespective of the site of sampling. It was concluded that the intercaruncular endometrium of cattle was far more active physiologically than recognized previously (Dhaliwal *et al.* 2002).

f. Endometrial Cytology

Uterine contamination elicits neutrophil migration from peripheral blood to the uterine lumen for phagocytosis of the contaminating organisms and happens to be the primary defense mechanism (Tizard, 1996). The presence of neutrophils in the lumen therefore, is an excellent indication of an active inflammatory process (Wade and Lewis, 1996). Once the offending contaminations are removed, inflammation subsides and neutrophils no longer migrate to the lumen. Therefore, in the normal uterus, examination of the luminal contents reveals no neutrophils (Klucinski *et al.* 1990b). There is a proportionate increase in the neutrophil with the increase in pathogen density. Prunner *et al.* (2014a) observed that uterine bacterial growth density increased from calving to 15 days postpartum and decreased from day 21 onwards; the corresponding PMN% in cytological samples decreased from calving to day 9, then surprisingly increased around days 15 to 21, and decreased thereafter. However, at each sampling period, the proportion of PMN was strongly depended on bacterial counts. This relationship of cellular response in active inflammation is the basis for cytological examination of the uterus (LeBlanc *et al.* 2002a).

The recovery of potentially pathogenic organisms by uterine culture might be erroneous due to contamination from the non-uterine sources (Zargham-Khan *et al.* 1998). Moreover, in chronic endometritis, physical examination including vaginoscopy, may not reveal an evidence of inflammation, such as hyperemia, exudates or pus (Mateus *et al.* 2002a; Sheldon, 2015; MSD Manual, 2023). Under both circumstances, as well as in the absence of any systemic signs of uterine infection in fresh postpartum cows, the cytological examination can be extremely helpful in establishing a diagnosis (Zerbe *et al.* 1996; LeBlanc *et al.* 2002 a,b). Cervical cytological examination in women (Glenthoj *et al.* 1986) and endometrial examination in mares (Wingfield-Digby, 1987), which has also been extended for the examination of the reproductive tract in cows (Mateus *et al.* 2002b; LeBlanc *et al.* 2002a; Sheldon *et al.* 2009a). In light of the aforesaid, Meira Jr *et al.* (2012) considered endometrial cytology as the gold standard for diagnosis of subclinical endometritis.

Methods of sample collection: Different methods have been used to assess the uterine cytology to diagnose endometritis and includes the use of (i) uterine biopsy (ii) guarded wet cotton swab, (iii) gentle aspiration of the uterine contents (iv) low-volume uterine lavage (v) cytobrush, and (vi) cytotape - techniques (Kasimanickam *et al.* 2004; Ahmadi *et al.* 2005 ; Gilbert *et al.* 2005; Kasimanickam *et al.* 2005b; Barlund *et al.* 2008; Sheldon *et al.* 2009a).

The limitations of uterine biopsy, as mentioned earlier restrict its practical application. The guarded wet cotton swab and gentle aspiration of the uterine contents have been undertaken in slaughter house material and are therefore less relevant for their field application (Ahmadi *et al.* 2005). Moreover, swabbing method caused increased epithelial collection, which restricts its usefulness. Low-volume uterine lavage, cytobrush and cytotape are therefore the most common methods utilized for endometrial cytology.

Gilbert *et al.* (2005) used the low-volume uterine flush to collect samples for endometrial cytology from cows between 40 and 60 days in milking. For the low-volume uterine lavage, an infusion pipette is protected with a sanitary chemise, which is punctured before the pipette is passed though the cervix. At any place in the uterus, 10 to 20 ml of sterile saline solution is infused, the uterus is then massaged for up to 5 to 10 seconds (Barlund *et al.* 2008) and a portion (2 to 5 ml) is harvested. A Folley catheter can also be used to perform a low volume lavage in a manner similar to embryo flushing (Galvao *et al.* 2010 a). After collection, the sample is to be centrifuged in a conventional or a cytospin centrifuge. If using a conventional centrifuge, most of the supernatant needs to be discarded and one drop of the remaining fluid is smeared onto a glass slide. For the cytospin, ~ 150 ìl of the collected sample is loaded in the cytospin container and centrifuged at 700 g for 5 min.

For the cytobrush, a Pap smear cytology brush is attached to a metal rod that is fitted through a metal pipe similar in diameter to an insemination pipette (Figure 3). The tool is protected with a plastic or metallic sheath protector during insertion into the vagina, and then is exposed for passing through the cervix. At the uterine body, the cytobrush is exposed and the body wall is pressed slightly against the cytobrush while the cytobrush is rolled two or three times. After that, the tool is exteriorized, and the cytobrush is smeared onto a glass slide. Pascottini *et al.* (2015) modified the cytobrush technique and described a new method for sample collection that consisted of using a paper tape rolled around the top of an insemination catheter. This method reduced the contamination of the stained smears with erythrocytes and a better preserved structure of epithelial cells than when using cytobrush could be obtained. Moreover, this system also permitted to obtain a cytology sample prior to AI (Pascottini *et al.* 2017). Sampling during AI may be an ideal proposal in this perspective because it does not require extra manipulation of the animals and offers opportunities to determine the uterine health status at the moment the animal is fertilized.

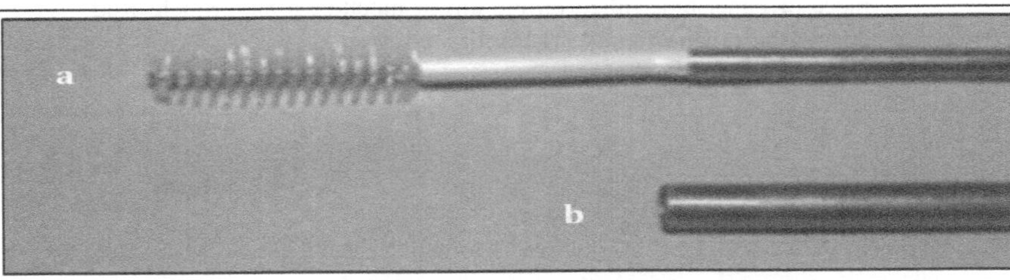

Figure 3: Components of a cytobrush assembly (a) Bristles in the cytobrush (b) Rod

Cytobrush is easily available in the Indian market. The limitations of cytobrush (endometrial damage and presence of excessive erythrocytes in the stained smear; Figure 4) and the cytotaping device (Pascottini *et al.* 2015; Figure 5 and Figure 6) (expensive and unavailable in open market), made us to develop a 0.4 cm thick stainless steel "bovine endometrial cytotoping device" which is efficient economical, safe, and autoclavable in providing reliable endometrial smears (Figure 7). The device, covered for protection from vaginal contamination with a bovine split universal sheath (IMV Technologies, Paris France), is introduced through cervix into t he uterine body. The sheath is removed when the cytotape device eaches external-os of the cervix. On reaching the uterus, the apical portion of the assembly, having a rolled over 2.0 cm wide sterile paper tape, is exposed into the uterus by

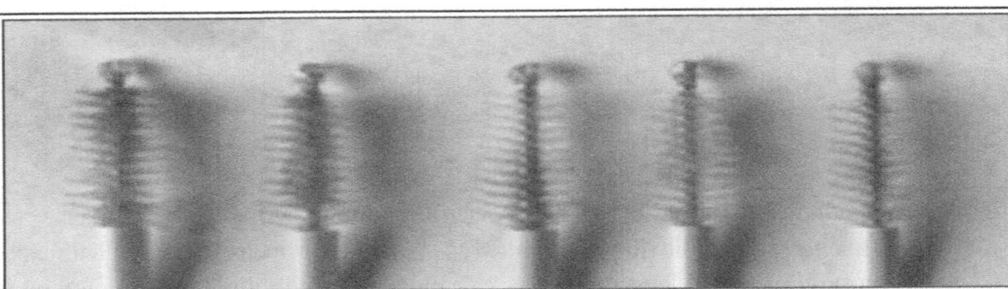

Figure 4: Varying amounts of blood stains on the cytobrush culminating into injury and/or RBC contamination

Figure 5: Components of cytotaping device (Pascotinni *et al.* 2015) (a) Cytobrush (b) Cytobrush adapted to stylette (c) Equine infusion pipette (d) A 1.5 cm piece of paper tape rolled on the top of pipette (e) Sani-Shield Rod® which covers the pipette finally (f) Configured assembly

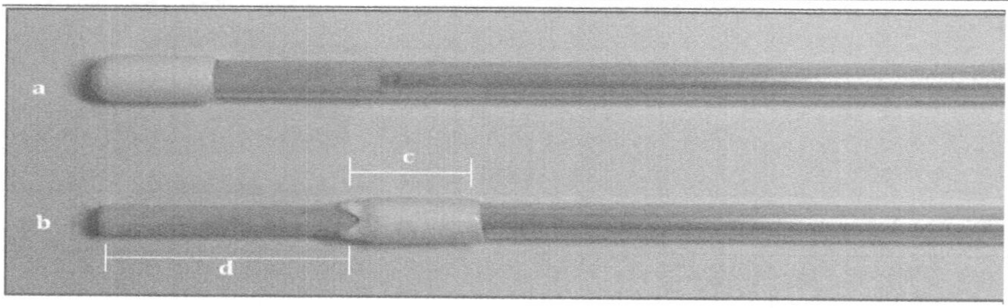

Figure 6: Apical components of cytotaping device (Pascotinni *et al.* 2015) (a) Sani-Shield Rod® covered pipette with paper tape, (b) Released cyto tape from Sani-Shield Rod® (c) Rubber top (d) Paper tape rolled on equine infusion pipette

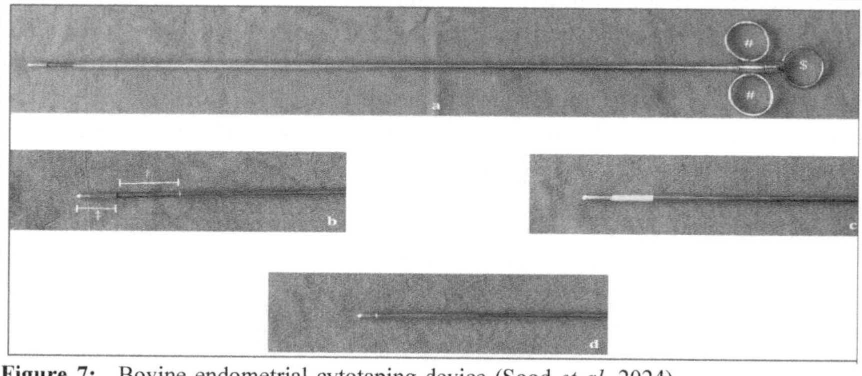

Figure 7: Bovine endometrial cytotaping device (Sood *et al.* 2024)
(a) circular metallic rings are marked as '# and $' (b) Open catheter with threaded portion of the stylette marked as '†', elongated and elliptical blob marked as '‡' (c) Threaded portion of the catheter rolled with cytotape (d) Closed catheter

pushing it from the other end of the device. On being exposed, the tape is rolled thrice against the dorsal endometrial surface of the uterine body. Thereafter, the apical portion is retracted into the device and removed from the cow. Outside the cow, the paper tape is rolled onto a customized microscopic glass slide of a width similar to that of the paper tape (Negi, 2022).

Irrespective of the method of cytologic sampling, smears need to be air-dried, stained, and microscopically evaluated. The staining method should be fast, easy to perform under practical circumstances, and yield high-quality samples allowing accurate interpretation (Pak *et al.* 1980; Yang and Alvarez, 1994). According to the literature, in more than 90% of the studies, a modified Wright-Giemsa staining, such as Diff-Quik, Tincion 15, Haema Quick, Hemacolor, is used (Gilbert *et al.* 2005; Kaufmann *et al.* 2009; Plöntzke *et al.* 2010 a,b; Melcher *et al.* 2014). Because these stainings are fast and easy to perform, they are well accepted and widely used, although they have not been compared with a gold standard to evaluate more objectively the PMNs-to-epithelial cells ratio. Naphthol-AS-D-chloroacetateesterase (CIAE) is an enzyme-histochemical method in which PMNs appear bright red after staining and is therefore regarded as the preferable staining to identify and count PMNs (Leder, 1970; Overbeck *et al.* 2013). Because a relatively small number of PMNs is needed to consider a cow positive for subclinical endometritis, it is essential to have a good quality staining not to miss any PMN and hence sub-evaluate the number of cows suffering from subclinical endometritis. Prepared smear must contain epithelial cells (to confirm correct site of collection); if no epithelial cells are seen, there is no assurance that the sample was taken from the uterus (Azawi, 2008).

After staining, all cells, including epithelial cells but excluding erythrocytes, are counted under the microscope, and the proportion of PMN out of a total of 200 cells is calculated (Figure 8).

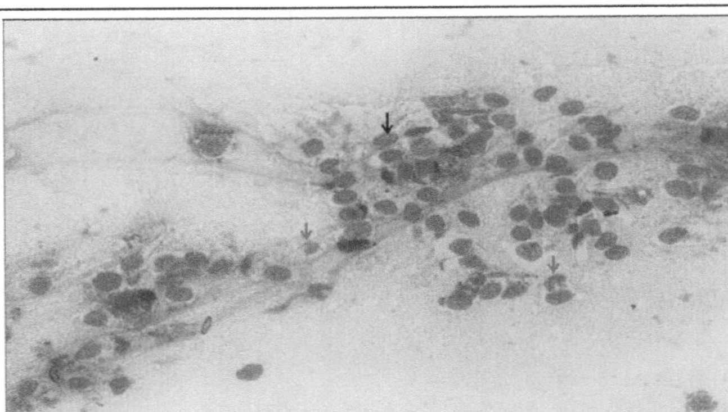

Figure 8: Components of endometrial smear (Red arrows = PMN cells; Black arrows = Endometrial cells)

With a little or no problem in detecting clinical uterine infection, subclinical endometritis detection on the other hand remains a serious challenge. Determination of uterine cytology remains one of the most promising methods to rule out subclinical endometritis in cows. The approaches for uterine cytology differ in the technique and cutoff points of PMNs at different days postpartum. Hence, there is lack of consensus in the cutoff percent of PMNs for establishing subclinical endometritis in cows at different days postpartum (Table 2).

Table 2: Reported prevalence of subclinical endometritis in some studies that used different sampling methods, postpartum diagnosis periods, and PMN% cutoff values.

Sampling method	**Postpartum diagnosis period**			**PMN%**
	Week 3–5	**Week 5–7**	**≥ 7 week**	
Cytobrush				
Lopdell *et al.* (2011)	35.0%			>18.0%
Kasimanickam *et al.* (2004)	35.1%			>18.0%
Heidarpour *et al.* (2012)	13.5%			>18.0%
Kaufmann *et al.* (2010a)	12.4%			>18.0%
Barrio *et al.* (2014)	17.6%			>18.0%
Madoz *et al.* (2013)	21.5%			>8.0%
Dubuc *et al.* (2010a)	19.3%			>6.0%
Plöntzke *et al.* (2010a)	38.0%			>5.0%
Lopdell *et al.* (2011)		7.0%		>18.0%
Kasimanickam *et al.* (2004)		34.0%		>10.0%
Barlund *et al.* (2008)		11.8%		>8.0%
Madoz *et al.* (2013)		16.0%		>6.0%
Plöntzke *et al.* (2010a)		19.0%		>5.0%
Barrio *et al.* (2014)		14.9%		>5.0%
Madoz *et al.* (2013)			16.0%	>4.0%
Dubuc *et al.* (2010a)			11.1%	>4.0%
Uterine lavage				
Hammon *et al.* (2006)	51.8%			>25.0%
Barlund *et al.* (2008)	15.8%			>8.0%
Gilbert *et al.* (2005)		53.0%		>5.0%
Cheong *et al.* (2011)		25.9%		>10.0%
Cytotape				
Pascottini (2016) (at AI, cows)			27.8%	≥1%
Pascottini (2016) (at AI, heifers)			7.86%	≥1%

In the postpartum cows, subclinical endometritis seems to be associated more with the postpartum recovery of the endometrium than with bacterial infection. Baranski *et al.* (2012) observed some controversies among authors in relation to the postpartum period from which a physiological endometrial inflammation should be considered as pathological subclinical endometritis. Therefore, depending on the sampling period after calving, different studies establish a different degree of PMNs infiltration as cutoff point to diagnose subclinical endometritis. Hence, the variation in both the days postpartum when samples are taken and the cutoff value applied to define an animal as subclinical endometritis positive, the prevalence of subclinical endometritis fluctuates widely among different studies (Baranski *et al.* 2012; de Boer *et al.* 2014; LeBlanc, 2014). Hence, prior to interpretation of the cytological samples, it is imperative to consider the postpartum stage at which the sampling is done. During the first month postpartum, healthy cows may show relatively high percentages of PMN in cytological samples, and therefore, diagnosis of subclinical endometritis during this period will be less accurate than at a later stage (Arias *et al.* 2018). Subclinical endometritis seems to be associated more with the postpartum recovery of the endometrium than with bacterial infection. In one of the related study (Baranski *et al.* 2012), using three different thresholds of PMN% at different postpartum days (>18% PMNs at first examination (4th week postpartum) and >10% at second examination (6th week postpartum) (2) > 8% in both exams, and (3) >5% in both exams indicated that the number of cows categorized as having endometritis increased as the threshold was lowered, and ranged from 18.9% to 75.4% . With all three thresholds, the number of cows categorized as having endometritis at the first examination was approximately double that at second examination so that 6.1% to 17.0% of the cows that were negative in the first exam were positive in the second, and 7.4% to 33.3% were positive in both exams. The cows were open for a significantly greater number of days if categorized as having endometritis with the first threshold at first examination, or with either the first or the second threshold at second examination.

Kasimanickam *et al.* (2005b) used receptive operative characteristic analysis to identify the PMN% above which fertility was significantly reduced, and this percentage was 18% for samples taken 20 to 33 days postpartum and 10% for those taken 34 to 47 days postpartum. Other authors also established the cutoff PMN% for diagnosing subclinical endometritis based on detrimental effects on subsequent reproductive performance (Barlund *et al.* 2008; Dubuc *et al.* 2010a; McDougall *et al.* 2011; Madoz *et al.* 2013), and some (Hammon *et al.* 2006; Plontzke *et al.* 2010a,b; Gilbert *et al.* 2005) used arbitrary values. In general, most authors used PMN% thresholds of 15 to 18% for subclinical endometritis diagnosis at 21 to 30 days postpartum and values of 4 to 10% for diagnosis at later periods. Prunner *et al.* (2014a) categorized clinically normal cows at 21 days postpartum as having subclinical endometritis when PMN% ≥5% and found that subclinical endometritis - positive cytological samples had an average of 30% PMN; however, on day 28, cows previously categorized as having healthy uteri (i.e., <5% PMN on day 21) had a similar PMN% as those categorized as having subclinical endometritis, and for both groups, it averaged 15%.

In contrast to the postpartum cyclic cows prior to resumption of regular cyclicity, the follicular phase of the estrous cycle evinced an increased infiltration of PMN in the endometrium elicited under estrogenic influence (Ohtani *et al.* 1993).

g. Uterine Lavage Sample Optical Density (ULSOD)

Clinical endometritis is defined as the presence of purulent or mucopurulent uterine exudates in the vagina, after 21 d postpartum, not accompanied by systemic signs (Sheldon *et al.* 2006).Although this definition is largely accepted and used by clinicians and researchers, Dubuc *et al.* (2010a) challenged assumptions of the latter method of diagnosis, showing that cows with purulent vaginal discharge did not always present endometrial inflammation. Hence, vaginal discharge can be erroneously associated with uterine inflammation when the discharge is a result of vaginitis or cervicitis. Moreover, some of the other techniques are prone to observer bias and variation between observers, and lack of training could lead to an erroneous diagnosis of clinical endometritis when scoring vaginal discharge by vaginoscopy.

Principle: Optical density measurement provides a numerical and objective value of the absorbed light in a sample, and it has been used to measure concentration of cells or proteins in samples (Glasel, 1995; Metris *et al.* 2006). ULSOD might be influenced by the exudates and cellular debris accumulated because of endothelium inflammation (Bondurant, 1999), and by the presence of bacteria inside the uterine lumen of cows affected with endometritis (Miller *et al.* 2007). With the low-volume uterine lavage technique, only the presence of a mucopurulent secretion inside the uterus is assessed, eliminating false-positive diagnoses that are more common with other methods of diagnosis such as rectal palpation and visual inspection of vaginal discharge (LeBlanc, 2008; Plontzke *et al.* 2010 b; Potter *et al.* 2010).

Procedure: A total of 20 mL of sterile saline solution is infused into the uterus and agitated gently, and a sample of the fluid is aspirated. The volume of recovered fluid ranges from 5 to 15 mL. All of the samples are visually scored by one investigator, who assesses the presence of a purulent or mucopurulent secretion in the uterine lavage sample. The score ranges from 0 to 2, with 0 indicating absence of a purulent or mucopurulent secretion in the lavage sample, 1 indicating a bloody but not purulent sample, and 2 the presence of pus in the lavage sample. Cows with a score of 2 were considered with clinical endometritis. The collected uterine lavage samples are processed for optical density measurement as follows: hypertonic saline solution (10%) was added to the uterine lavage sample in a 1:1 proportion, samples are then incubated at room temperature for 30 min (if frozen) and subsequently centrifuged for 30 min at 3,000 × g at 4°C. The supernatant is collected and an aliquot of 0.2 mL is added to a 96-well microplate and optical density is assessed in a microplate reader and is measured at 620 nm as the said wavelength presented the best receptive operative characteristics curve to detect clinical endometritis (Machado *et al.* 2012a).

Interpretation: The recommended threshold for $ULSOD_{620}$ was 0.059, where the sensitivity and specificity were 100 and 82.2%, respectively in cows that were 34 to 36 days postpartum. Cows with $ULSOD_{620}$ d"0.058 were 1.21 times more likely to conceive than cows with $ULSOD_{620}$ >0.058; moreover, the median calving-to-conception interval for cows that had ULSOD620 d"0.058 was 122 d compared with 148 d for cows that had $ULSOD_{620}$ >0.058. Cows that were positive for *Arcanobacterium pyogenes*, diagnosed with metritis, or had retained placenta had 4.0, 1.4, and 1.7 times higher odds of having $ULSOD_{620}$ >0.058, respectively. Cows with $ULSOD_{620}$ >0.058 had a higher percentage of neutrophils in the uterine lavage samples than cows with $ULSOD_{620}$ d"0.058. ULSOD measured at 620 nm can be used as an objective indicator of uterine health in dairy cows, principally for clinical endometritis (Machado *et al.* 2012a).

h. Ultrasound

The ultrasound use in detection of uterine infections is based on the persistence or presence of uterine fluid after a certain stage postpartum, which reflects the inflammation or infection mediated effects on the uterine lining and impaired uterine involution (Lenz *et al.* 2007; Oral *et al.* 2009; Purohit *et al.* 2013). Presence of uterine fluid or the cervical and uterine diameters at different postpartum stages have therefore been used to depict uterine infection in cows. An *inter se* relationship between the two has also been reported. Bovine reproductive organs are most commonly scanned per-rectum using a linear array trans-rectal transducer. Linear array transducers of 5.0 and 7.5MHz frequency ranges are most commonly used in cattle to perform reproductive ultrasound examination (Fricke, 2002). Mateus *et al.* (2002a) along with some others (Kasimanickam *et al.* 2004; Barlund *et al.* 2008; Oral *et al.* 2009) noted that ultrasound uterine measurement was easy, quick and useful technique allowing reliable comparison of results. Okano and Tomizuka (1987) and Sheldon *et al.* (2006) reported that ultrasound examination is an objective, noninvasive, practical, and exact method. The latter two studies agree with such a statement when the diagnosis can be made in real time taking exact measures.

Healthy cows achieve a uterine horn diameter of 3 to 4 cm by 25 to 30 days post partum, and cervical diameter of <5 cm by 40 days post partum, but involution of the uterus and cervix is not complete until approximately 40 to 50 days post partum (Gier and Marrion, 1968; Sheldon, 2004). After birth, the uterus is enlarged; uterine macroscopic involution occurs 3 to 5 weeks postpartum, when a healthy uterine horn must have a diameter smaller than 5 cm (Gier and Marion, 1968). Slama *et al.* (1991) proved that uterine infections predispose to a delayed uterine involution. LeBlanc *et al.* (2002a) concluded that the sizes of the uterus and cervix in cows with clinical endometritis were associated with a decrease in pregnancy rate if the size of the uterus was >8 cm in diameter and the cervix was >7.5 cm in diameter in cows between 20 and 33 days in milk. In the latter study, the uterine and cervical sizes at 20 to 33 days postpartum and 34 to 47 days postpartum were not associated with relative pregnancy rate. Furthermore, it was noted that higher proportion of cows had cervical and uterine diameters that differed from the previous study (Meira Jr. *et al.* 2012). Accordingly, transrectal ultrasonography and endometrial cytology were suggested for the definitive diagnosis of subclinical endometritis (Lenz *et al.* 2007; Oral *et al.* 2009; Melcher *et al.* 2014).

Apart from quantum of bacterial infection, few other factors affecting uterine involution include age, breed and nutrition so that delayed uterine involution is not a specific indicator of uterine disease (Fonseca *et al.* 1983). Due to the aforesaid reasons causing a variation in findings of cervical and uterine diameters, detection of uterine fluid may be more reasonable to diagnose uterine infections. Ultrasonographic intrauterine fluid determination after week 3 postpartum has been a reliable diagnostic method for endometritis, with good sensitivity and specificity (MSD Manual, 2023). Presence of ≥ 2 mm of fluid in the uterine lumen after ultrasonographic visualization leads to sample collection using the cytobrush technique and microscopic assessments to determine the percentage of PMNs (Jaureguiberry *et al.* 2017). The presence of uterine fluid is more marked in cows with severe rather than mild infection. Kasimanickam *et al.* (2004) concluded the intrauterine fluid volume identified using ultrasonography was significantly correlated with uterine swab bacterial growth, and that infection significantly impaired uterine involution. Although the uterine body diameter reduced to pre-gravid size by 6-

week postpartum, the intrauterine fluid volume remained higher in cows with severe puerperal infection compared to control and mildly infected cows.

i. Uterine pH

The pH of the uterine lumen differs with the stages of the estrous cycle and may be used in assessing the degree of bacterial contamination of the uterine lumen. In the estrogen dominance, uterine pH becomes higher than in progesterone dominance (Fukusima, 1972). The pH of vaginal mucus turns to alkaline in infected cows (Fukusima, 1972). The uterine pH was lowest 2 days prior to ovulation, and at estrus, the value is 7.30 (Mather, 1975). Another study reports that in normal breeding cows, the pH of the vaginal mucus is 7.2 ± 1.1 compared to 8.2 to 8.8 in cows with endometritis and metritis (Boitor *et al.* 1980). May (1996) found that the pH of discharge collected from field cases of endometritis varied from 6.9 to 7.3, which favored the growth of *A. pyogenes*.

An impediment in the determination of uterine pH is the lack of the development of special sensing probes that can be introduced directly into the bovine uterus.

j. Whiteside Test

The test was devised by Whiteside in 1939 and was modified by Schalm and coworkers in 1971. The earliest reports citing use of Whiteside test as a routine cow side test to diagnose the subclinical endometritis was way back by Popov (1969) and Polyantsev and Popov (1972).

Principle: On heating of genital discharge with 5% NaOH upto boiling point, strands of genetic material of neutrophils react with NaOH to produce yellow color. Gupta *et al.* (2011) proposed that the underlying mechanism of Whiteside Test is a result of reaction of sodium salts with the nucleic acids of white blood cells.

Method: For conducting Whiteside test, 1 ml of estrual cervical mucus is heated with equal volume of 5 to 10% sodium hydroxide solution (Merck, Mumbai, India) up to boiling point and after cooling under running tap water there may be a change in the color and its intensity. The reaction is considered positive when the color turns yellow. The grading of color vis-à-vis severity of infection is normal (turbid or no color), mild infection (light yellow color), moderate infection (yellow color) and severe infection (dark yellow color). In other words a proportionate increase in number of leucocytes with an increase in uterine infection intensifies the yellow color (Figure 9). This test has the greatest advantage of establishing the degrees of infection in just a few minutes.

Figure 9: Change in color of genital discharge after Whiteside test (test tube on extreme left shows mild yellow discoloration; test tube in the middle shows intense yellow discoloration; test tube on extreme right shows no change)

However, there is some inconsistency in the results of this test, which restricts its validation (Purohit, 2008; Raja *et al.* 2012). Presence of the Whiteside reaction in normal and fertile cows is one of the potential reasons.

The results of Whiteside test showed 15% of healthy and fertile cows revealing no bacterial growth exhibited a positive Whiteside reaction. In contrast, 71.25% samples were positive for bacterial culture, whereas 75.0% showed Whiteside reaction. The study recommended the use of Whiteside test as a prime modality for ascertaining nonspecific bacterial infection of repeat breeding cattle before subjecting them to any therapy (Bhat *et al.* 2014). Deori (2002) found equal percentage of cows with moderate and subclinical endometritis using Whiteside test in cervico-vaginal mucus. In cycling cows, use of Whiteside test can be helpful in diagnosing subclinical endometritis. Savitha *et al.* (2005) found that 30% of cows and heifers which were diagnosed by clinicogynaecological examination as suitable for AI were actually suffering from subclinical endometritis.

The discrepancy between the bacterial culture and Whiteside test reaction can be due to evaluation of genital discharge from cows with negative microbial culture, but are into metestrus, especially those with metestrual bleeding.

k. Leucocyte Esterase

Principle: Leukocyte esterase is an esterase-like enzyme produced by leukocytes and is mainly used as a marker to detect the presence of leukocytes in a sample. The enzyme hydrolyzes an acid ester to release an alcohol. The alcohol reacts with a daizonium salt to form a purple product. The amount of color change in the dye is proportional to the number of white cells present in the sample. The dye reaction requires time to become visible. There is also a maximum color change that can be obtained with the strip reagent. This is commonly set at approximately 5-15 cells/hpf. A positive **leukocyte esterase test** is often indicative of infection. It has been used to detect infections of urine, amniotic fluid (humans) and reproductive tract in cattle (Figure 10).

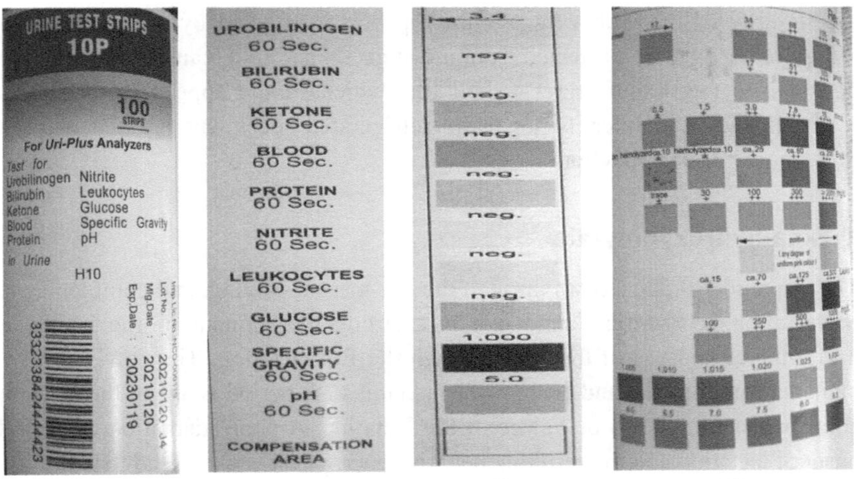

(A) (B) (C) (D)

Figure 10: Urinary test strips (A); Tests that can be performed and reading time (B); Color scheme depicting different grades of outcome for different tests (C and D)

The test is usually performed using a test strip or an automated analyzer. Both methods are widely used and can provide reliable results when used correctly. The test has been utilized more commonly to detect urine infection, whereby it is important to note that the **leukocyte esterase test** is not a definitive diagnostic tool and should be used in conjunction with other tests (such as the nitrite test for urine).

The test has been proposed as a cow-side test for diagnosing postpartum subclinical endometritis (Santos *et al.* 2006; Cheong *et al.* 2012; Couto *et al.* 2013; Nazhat *et al.* 2018). The samples can be uterine lavage fluid or uterine cytobrush samples (Herath *et al.* 2007; Shimizu *et al.* 2012). The latter can be immersed in saline solution during 30 sec and then the strip introduced in the solution for 2 sec analysis (Arias *et al.* 2018). One drop of uterine lavage sample is added on the Multistix 10 SG (Bayer Corporation) reagent strip and the results can be interpreted 2 min.

Leukocyte esterase test showed a variable correlation with cytology. Santos *et al.* (2006) reported a sensitivity of 96% and specificity of 98%. However, Cheong *et al.* (2012) in a larger data set using >10% neutrophils as cut-off observed a sensitivity and specificity of 77% and 52%, respectively. It was concluded that the test is qualitative (Arias *et al.* 2018) and needs further improvement for its accuracy in diagnosing endometritis.

Precautions in using dipsticks say that alike all reagents stored in a dried form, the stability of the leukocyte esterase dipstick is affected by moisture and has therefore needs to be protected from moisture. In addition, careful attention must be given to the manufacturer's shelf life, which should appear on the label of every dipstick container. If not fresh, refrigerated specimens are acceptable up to 24 hours after collection.

I. Beta-hydroxybutyrate

The transition of dairy cows through the dry-period, calving, and commencement of lactation is a critical time and is a strong determinant of the health and performance success of the cow through the full lactation. The cow's ability to manage energy intake and demand through this time period is one of the important contributors to success or failure of the lactation. A poor adaptation to both the onset of lactation and the resulting negative energy balance can lead to a multitude of problems including clinical disease and impaired milk production (Duffield *et al.* 2009). One indicator of a poor adaptive response is elevations in circulating ketone body concentrations (Herdt, 2000). Whereas ketone elevations postcalving are part of a normal adaptive response, excess ketone accumulation is abnormal. Beta-hydroxybutyrate is the predominant circulating ketone body in ruminants, but there is a strong correlation between the whole blood concentrations of beta-hydroxybutyrate and acetoacetate (Kauppinen, 1983). Conceivably, subclinical ketosis may start at levels of beta-hydroxybutyrate greater than 1,000 ìmol/L; however, the decision to set an appropriate subclinical threshold using serum or plasma BHBA appears to be somewhat arbitrary. Kelly (1977) suggested that 1,000 ìmol/L be used to separate cows with low and high BHBA concentrations. Nielen *et al.* (1994) selected a subclinical ketosis threshold of 1,200 ìmol/L of BHBA based on the data appearing to be skewed around that value.

The probability of pregnancy was reduced by 20% in cows diagnosed subclinically ketotic in either the first or second week postpartum. In addition, elevated serum beta-hydroxybutyric acid concentrations in the first 2 weeks postpartum indicate an increased risk of metritis (Duffield *et al.* 2009), which is a usual precursor of the development of clinical and subclinical endometritis (Walsh *et al.* 2007). The most likely explanation of this finding is an indirect effect of elevated beta-hydroxybutyrate on metritis risk mediated through impacts of hyperketonemia on immune function (Suriyasathaporn *et al.* 2000). However, decreases in dry matter intake have been identified several weeks precalving in cows subsequently diagnosed with metritis postcalving (Huzzey *et al.* 2007). Less dry matter intake precalving likely results in greater non-esterified fatty acids precalving and consequently greater beta-hydroxybutyrate concentrations post-calving (Osborne, 2003).

m. Myeloperoxidase

Apart from leukocyte esterase, myeloperoxidase is another important enzyme secreted by leukocytes. Myeloperoxidase is a member of the mammalian peroxidase family, and is known to be effective in killing microorganisms (Guven *et al.* 2013). Parrilla-Hernandez *et al.* (2014) determined myeloperoxidase levels in uterine lavage samples collected from mares, and proved that increased myeloperoxidase concentrations were related to the occurrence of endometritis. In cows, however, no association was found between endometrial myeloperoxidase (in cytology samples) levels of 28 to 60 ng/ml and endometritis, which was in contrast to a much stronger association between endometrial leukocyte esterase and endometritis (Nazhat *et al.* 2018).

n. Immunological and Inflammatory Biomarkers

Different components of the pro-inflammatory, inflammatory and host generated immune response have been utilized for prediction, diagnostic and prognostic markers in relation to uterine infections in cows (Huzzey *et al.* 2011; Smith *et al.* 1998b; Hirvonen *et al.* 1999; Jeremejeva *et al.* 2012).

Uterine tissue based

Because cows that develop endometritis had a decreased influx of PMN into the uterus around calving and increased numbers of PMN later in lactation, which were associated with chronic bacterial infection, it was hypothesized that uterine tissue from cows diagnosed with endometritis would have decreased endometrial gene expression of pro-inflammatory cytokines around calving and increased endometrial gene expression of pro-inflammatory cytokines around the time of diagnosis of endometritis. Also, because pro-inflammatory cytokines (TNF-α and IL-1β) would induce the production of IL-8 by the uterine vascular endothelium, it was hypothesized that IL-8 would be decreased early in lactation and increased later in lactation in cows that develop endometritis (Szenci, 2016).

In addition to PMN mount in uterus, the neutrophil function is also decreased around the time of calving in high-producing dairy cows (Kehrli *et al.* 1989; Gilbert *et al.* 1993; Goff and Horst, 1997), especially in those that develop uterine disease (Cai *et al* 1994; Kim *et al.* 2005; Hammon *et al.* 2006). The impairment in leukocyte function around calving is thought to arise from a deficit in energy, a deficiency in vitamins and minerals, and an increase in estradiol and glucocorticoids (Goff and Horst, 1997). Therefore, it appears that not only individual leukocytes can be affected during transition into lactation, but also other cells with immune capabilities, e.g., endometrial cells (Davies *et al.* 2008). Lately, attention has turned to the role of the uterine endometrium in mounting an inflammatory response and modulating inflammation in cows with endometritis (Chapwanya *et al.* 2009; Gabler *et al.* 2009; Fischer *et al.* 2010). A lower local level of expression of pro-inflammatory cytokines in the endometrium soon after calving might impair activation of inflammation and clearance of bacteria, and lead to development of endometritis. Expression of the TNF-α gene in uterine tissue was decreased in cows with endometritis compared to control cows at calving and at Week 1. Similarly, IL-1β gene expression tended to be decreased in cows with endometritis compared to control cows at week 1, but tended to be increased at weeks 5 and 7. Cows with endometritis had increased IL-6 gene expression at calving and at week 7 compared to control cows. The IL-8 gene expression was increased in endometritic cows compared to control cows at week 7. Uterine disease was not significantly associated with IL-10 gene expression. A suboptimal cytokine expression could be due to an intrinsic defect in endometrial cell function, as has been observed with leukocytes (Nino-Soto *et al.* 2008), or extrinsic mechanisms affecting endometrial cell activity, such as a greater degree of negative energy balance in cows with endometritis (Hammon *et al.* 2006), or related to impaired function of infiltrating immune cells. Because differences in gene expression between cows that developed endometritis and control cows occurred mainly when negative energy balance was more severe (Beam and Butler, 1998), perhaps the state of negative energy balance affects the ability of endometritis cows to up-regulate gene expression of pro-inflammatory cytokines. The latter was supported by another study that compared the cows with severe negative energy balance with cows in moderate negative energy balance. It was observed that cows in severe negative energy balance had increased uterine pro-inflammatory cytokine gene expression at week 2 after calving compared to moderate negative energy balance cows (Wathes *et al.* 2009). It was inferred that the ability to clear infection was impaired in cows in severe negative energy balance because inflammation was still evident (increased pro-inflammatory cytokine gene expression) at week 2 after calving. Hence, in cows that developed endometritis, more severe negative energy balance may actually impair the ability to up-regulate pro-inflammatory cytokines in the first week of lactation; however, after the negative energy balance was resolved later in lactation, endometritis cows mounted an inflammatory response.

Galvao *et al.* (2011) compared the gene expression of important pro-inflammatory (TNFα, IL-1β, IL-6; all three in uterine tissue) and anti-inflammatory (IL-10; in uterine tissue) cytokines, and the main neutrophil chemokine (IL-8; in serum), from calving to week 7 after calving between the cows that developed endometritis (>10% neutrophils) and those that were healthy. Expression of the TNFα gene in uterine tissue was decreased in endometritis cows compared to control cows at calving and at Week 1 postpartum. Interleukin-1β gene expression tended to decrease in cows with endometritis compared to control cows at Week 1 postpartum, but tended to be increase at weeks 5 and 7 postpartum. Cows with endometritis had increased IL-6 gene expression at calving and at week 7 postpartum compared to control cows. IL-8 gene expression was increased in endometritic compared to control cows at Week 7 postpartum. It was inferred that the uterine disease was not significantly associated with IL-10 gene expression, as was found earlier by Nino-Soto *et al.* (2008).

Loyi *et al.* (2013) observed a significantly higher expression of various inflammatory mediators viz. CD-14, TLR-4, IL-1β, IL-6, IL-8 and TNF-α mRNA in the endometrium of buffaloes with clinical endometritis, whereas significant up-regulation of CD-14 (1 to 2-fold), IL-6 (15 to 36-fold), IL-8 (8 to 14-fold) and TNF-α (10 to 11-fold) mRNA was observed in subclinical endometritis; the samples collected from abattoir specimens. Similarly, in another study on cattle it was observed that the expression of endometrial transcripts of pathogen recognition receptors and major proinflammatory cytokines is dependent on the severity of inflammation and stage of cyclicity (Loyi *et al.* 2015). Enhanced expression of CD-14 and IL-8 at the follicular stage and IL-6 and TNF-α in luteal stage could be used as a possible target gene for developing suitable diagnostic markers in cows with subclinical endometritis in future studies.

Ishikawa *et al.* (2004) investigated the correlation between the IL-6 concentration and the occurrence of postpartum diseases and found that cows suffered with endometritis had a higher level of pre-partum IL-6 than control. They suggested that alteration in the IL-6 concentration during pregnancy was one the useful tool for predicting postpartum diseases in dairy cattle.

Kim *et al.* (2005) reported that dairy cows suffered with endometritis during the 3rd and 4th week postpartum period had a higher concentration of TNF-α than normal cows. Islama *et al.* (2013) observed the significantly higher concentration of IL-10 in clinical metritis than normal cows at 15 days before calving, at calving and 15 days postpartum. However, the significant difference was found only at 30 d postpartum for clinical endometritis cows.

In another study, Kasimanickam *et al.* (2013) found that cows with metritis or clinical endometritis had higher serum concentrations of IL-1β, TNF-α, and IL-6 compared to normal cows and suggested that loss of body conditions mediated increases in cytokines and thereby prolonged the uterine inflammation in dairy cows.

Gabler *et al.* (2010) revealed time-related expression of inflammatory cytokines and acute phase proteins, with significant peak expression on the day 17 postpartum period as a possible mucosal immune response in the uterus. In addition, they found that IL-1β, IL-8, and haptoglobin mRNA expression were correlated significantly with the proportion of PMN. Collectively, it suggests that the dynamics of these acute phase proteins differs between normal and inflamed uterus and it needs further investigation. Further, the moderate level of expression of α1-acid glycoprotein in the healthy uterus has been reported by some researchers (Lecchi *et al.* 2009; Rahman and Lecchi, 2010), but there is no information on acid glycoprotein expression in the infected uterus. Fischer *et al.* (2010) found that the expression of IL-1β, IL-8, and TNF-α mRNA was significantly higher in cows with subclinical or clinical endometritis compared with healthy cows, while there was no indication or correlation with uterine health for IL-6 and haptoglobin (Hp) transcripts. They suggested that IL-1β, IL-8, and TNF-α might represent potential marker genes for the detection of cows with subclinical endometritis and for monitoring new therapeutic approaches. Ghasemi *et al.* (2012) found 20, 30, and more than 50-fold higher mRNA expression of TNF-α, IL-6, and IL-8 level, respectively, in subclinical endometritis than healthy cows. They suggested that IL-8 gene expression might be useful to predict endometrial inflammation.

Galvao *et al.* (2012) found that *E. coli*-stimulated monocytes from cows with metritis had lower expression of TNF-α, IL-1β, and IL-6 than healthy counterpart while there were no significant differences in IL-8 or IL-10 expression in these cows. Collectively, it suggested that pro-inflammatory cytokines expressions are differentially altered during postpartum uterine infections. Chapwanya *et al.* (2013) found that expression of IL-1β, TNF-α, and serum amyloid A3 (SAA3) genes were increased by 121-, 357-, and 721-fold, respectively, during 6 h post *E. coli* stimulation. However, IL-1β, IL-6, IL-8, and TNF-α gene expression was decreased, whereas SAA3 expression was further increased to 3452-fold after 24 h of *E. coli* stimulation compared to 6 h. They suggested that better understanding of localized endometrial expression of SAA3 was needed to use SAA as a sensitive diagnostic marker for *E. coli* infection in cattle

Plasma /serum based

The other most intensively researched acute phase proteins in connection to inflammation are haptoglobin (Hp), SAA, and α1- acid glycoprotein (AGP). Hp and SAA show significant positive correlations in serum concentration values and increase 100-fold as opposed to the three to four fold increase of AGP in dairy cattle (Eckersall *et al.* 2001). SAA is a rapidly reacting acute phase protein, which shows a high level of inflammation in cattle. For example, it has been stated that the mean of SAA levels in cattle with acute diffuse peritonitis is 312.4 µg/ml, and that a statistically significant increase can be useful in clinical medicine (Ansari-Lari *et al.* 2008). Especially in the diagnosis of mastitis, the determination of SAA levels in blood serum and milk has been reported to be useful in identifying acute, chronic and subclinical conditions of the disease, as well as in distinguishing between mild and moderate conditions (Eckersall *et al.* 2001; Gronlund *et al.* 2005). In a study by (Basbug *et al.* 2020), SAA levels in healthy cattle were consistent with previous studies. The 4 to 9 fold increase in SAA in cows with arthritis, pneumonia, mastitis, retained placenta compared with healthy cattle, were similar in magnitude to that reported previously in cows with traumatic reticuloperitonitis, mastitis, metritis, pododermatitis, and abdominal infection (Hirvonen *et al.* 1998; Nazifi *et al.* 2008).

Choosing amongst Hp and SAA, Hp has been most extensively investigated in relation to uterine health in dairy cattle (Peterson *et al.* 2004).

Haptoglobin: Hp is a plasma macromolecule (>1000kD) (Eckersall and Conner, 1990) with beta-alpha-alpha-beta chains connected by disulfide bridges (Morimastsu *et al.* 1991). It is produced by neutrophils, liver hepatocytes, and alveolar cells (Lim *et al.* 1998; Levy *et al.* 2010; Aronowski and Zhao, 2011). Production of Hp by macrophages and neutrophils (Levy *et al.* 2010; Theilgaard-Monch *et al.* 2006) increases with hemolysis and leukocytosis (Rother *et al.* 2005).

Cooray *et al.* (2007) demonstrated elevated Hp of multiple isoforms in circulating granulocytes of healthy cattle. The main function of Hp is to scavenge free hemoglobin (Hb) during an oxidative stress (Langlois and Delanghe, 1996; Kristiansen *et al.* 2001). During infection, bacteria use the heme structure of Hb for growth and colonization (Hiss *et al.* 2004). The Hb is released by the rupture of erythrocytes (Rother *et al.* 2005). The released Hb contains an iron intermediate that reacts with downstream hydrogen peroxide resulting in formation of hydrogen radicals (Banerjee *et al.* 2012). The Hb iron intermediate has oxidative potential (Banerjee *et al.* 2012) that causes free heme to be released into circulation and prevent clearance from the body (Buehler *et al.* 2009). Hp complexes with Hb in an irreversible form (Rother *et al.* 2005) to reduce the oxidative potential, thereby limiting iron availability to bacteria. Therefore Hp acts as an antioxidant (Yang *et al.* 2003; Tseng *et al.* 2004). Oxygen has shown an increased affinity for the Hp-Hb complex when compared to uncomplexed Hb (Banerjee *et al.* 2012). The Hp-Hb complex is considered bacteriostatic and antinflammatory (Levy *et al.* 2010). The Hp-Hb complex reduces the toxic potential of Hb, allowing its removal by the glycoprotein CD-163 (Freeman *et al.* 1990), a transmembrane protein found on monocyte and macrophage cells. Expression of CD-163 *in vitro* was influenced by administration of cytokine and TLR ligand. Down regulation of CD163 was achieved by administration of IFN-ã, TNF-α, IL-4, and lipopolysaccharide with glucocorticoids, IL-6, whereas IL-10 induced up regulation of CD163 (Van Gorp *et al.* 2010; Etzerodt and Moestrup, 2013). With nine scavenger receptor cysteine rich domains (Van Gorp *et al.* 2010), the third domain of CD163 is essential in Ca2+ dependent ligand formation with the Hp-Hb complex (Madsen *et al.* 2004). CD163 can also, uptake free Hb allowing a backup system if Hp is depleted or there is excessive hemolysis (Schaer *et al.* 2006). The uptake of free Hb is shown by high affinity CD163 subunits (Etzerodt and Moestrup, 2013). The binding of Hb to CD163 (uncomplexed or complexed with Hp) causes secretion of IL-6 and IL-10 as well as heme oxygenase, a stress response protein (Philippidis *et al.* 2004). If uncomplexed, Hb accumulates in the kidney. Therefore, Hp protects from renal failure due to the buildup of Hb in the kidneys as well as preventing of iron loss (Kim *et al.* 1998).

Hp has been measured in milk as well as serum. The early increase of Hp in milk confirms the hypothesis that Hp synthesis occurs in the mammary gland. Eckersall *et al.* (2001) showed a significant correlation between Hp in serum and milk and can increase markedly in cows with mastitis as well (Nielsen *et al.* 2004). The production of cytokines by macrophages suggests that acute phase proteins, specifically Hp can be produced in several tissues like the kidneys, gastrointestinal tract, muscle, alveolar cells of the mammary gland (Van Gorp *et al.* 2010), lung (D'Armiento *et al.* 1997), and oviduct (Lavery *et al.* 2004). These tissues are highly vascularized and contain extensive capillary beds to maximize exchange of blood from mammary arteries to capillaries, draining into interstitial tissues. The increased blood flow also increases circulating leukocytes and macrophages in the tissues. Tissue inflammation results in increased cytokine and lysosomal activity in epithelial cells and production of arachidonic acid from prostaglandins. These factors expedite transfer of leukocytes and acute phase proteins into the tissue.

Hp has been shown in numerous cases to be an inflammatory marker for cattle (Lipperheide *et al.* 1997; Nakagawa *et al.* 1997; Eckersall *et al.* 2001; Sheldon *et al.* 2001; Peterson *et al.* 2004; Ceron *et al.* 2005) with concentrations increasing during bacterial challenges (Petersen *et al.* 2004; Lai *et al.* 2007). Alsemgeest *et al.* (1995) observed an increase in serum Hp in acute, subacute, and chronic inflammations such as pneumonia, peritonitis, and abscess, which led to the use of Hp as a factor for diagnosing sick cattle (Skinner *et al.* 1991).

Cairoli *et al.* (2006) found that the concentrations of Hp and AGP fluctuated at the time of calving and in cows affected with postpartum endometritis. However, an elevation of plasma Hp concentrations in metritic cows could be found only in the most severely affected cows (Hirvonen *et al.* 1999) that revealed serum Hp concentration of 1.06 to 1.90 g/L, thereby suggesting an indication of acute infectious process in such cows (Drillich *et al.* 2007; Huzzey *et al.* 2009). This was also supported by Sheldon *et al.* (2001), who found that acute phase protein concentrations were increased by bacterial contamination. Reported sensitivities for the detection of metritis (50 to 79%) and specificities (54 to 87%), however, were only moderate (Huzzey *et al.* 2009; Dubuc *et al.* 2010a; Burfeind *et al.* 2014b).

Serum Hp levels significantly increase post parturition (Uchida *et al.* 2006; Gymnich *et al.* 2003). Hiss *et al.* (2009) showed the concentration of serum Hp was low (0.19±0.03 mg/mL) four weeks prepartum, increased to 1.6+0.2 mg/mL one week postpartum and returned to basal levels three weeks postpartum. Highest concentrations were recorded one week postpartum and declined to basal levels six weeks postpartum (Hiss *et al.* 2009). Hp levels during uterine infection and uterine involution have been correlated with illness status (Sheldon *et al.* 2001), as well as in cases of metritis (Smith *et al.* 1998 a; Chan *et al.* 2010). Huzzey *et al.* (2009) showed that cows with Hp serum levels higher than 1mg/mL three days postpartum were predicted to be 6.7 times more likely to develop metritis. Williams *et al.* (2007) observed that cows with high uterine pathogen growth density had higher peripheral concentrations of AGP, SAA, and Hp compared to low uterine pathogen growth density cows during 7 and 14 days postpartum period. A positive correlation between the three acute phase proteins and severity of disease and the extent of the tissue damage was also reported by Baumann and Gauldie (1994). In contrast, Pyorala (2000) revealed that the AGP levels did not correlate with the severity of disease compared to Hp, and thus the capacity of AGP in differentiating uterine infections was poor.

In dairy cattle, increase in Hp serum concentration has been implicated with decreased reproductive efficiency (Huzzey *et al.* 2015). Nightingale *et al.* (2015) demonstrated that increasing Hp concentration two to eight days postpartum increased days to conception. This study proposed that increase in the acute phase response, triggered by upregulation of innate immune responses, suppressed mitogen induced IFN-ã secretion negatively impacting reproduction. Plasma Hp levels one to three weeks postpartum also increased with bacterial contamination in the uterus (Sheldon *et al.* 2001). Galvao *et al.* (2010 b) demonstrated that cows with clinical metritis had elevated plasma Hp levels 7 days in milk and returned to baseline by 14 days in milk. In the same study, cows with subclinical metritis did not show elevated plasma Hp. Evidence of decreased hepatic function was shown in cattle with a negative energy balance, increased bilirubin levels, and increased plasma Hp concentrations (Bertoni *et al.* 2008). This study also showed that cows in this group exhibited lower fertility with greater number of days open, greater services per conception, and lower than average milk yield at 28 days postpartum.

The lack of specificity of Hp as a biological marker comes with advantages and disadvantages. The broad spectrum of stressors resulting in the acute phase response cause an unspecified increase in Hp 50 to 100 times the basal levels (Kim *et al.* 2011). Humblet *et al.* (2006) suggested that Hp and SAA were good markers for identification of healthy animals, but their ability to identify the diseased animals was low as most of the postpartum diseases were of chronic in nature. Several researchers observed that Hp concentrations are often undetectable in healthy cattle (Eckersall and Conner, 1988), but increase to about 50±100 times during an acute phase reaction, makes it the most prominent acute phase protein in cattle (Alsemgeest *et al.* 1994). On the other hand, SAA is a low constitutive acute phase protein in cattle increasing around 2±5 times or 10-fold during an acute phase reaction (Gruys *et al.* 1993; Alsemgeest *et al.* 1994). Nevertheless, SAA seems to react faster than Hp against acute phase reaction (Horadagoda *et al.* 1994), and thus SAA is considered as good marker for acute clinical conditions (Horadagoda *et al.* 1999), while Hp as a better marker for chronic conditions (Pyorala, 2000). Inspite of the higher Hp levels in diseased animals, a wide variation in the serum concentration of Hp in healthy and different disease subjects has been recorded (Table 3).

Table 3: Experimentally determined bovine haptoglobin standards

Status of animal	Serum concentration (mg/mL)	Reference
Healthy	0.0003	Weinkauf *et al.* 2005
Healthy	0.108±0.017	Lipperheide *et al.* 1997Nakagawa *et al.* 1997
Healthy	<0.259	Slocombe and Colditz, 2005
Healthy	0.35	Horadagoda *et al.* 1999
Healthy	0.108 ± 0.017	Khoshvaghti *et al.* 2009
Healthy	0.02-0.1	Eckersall *et al.* 2001
Acute Metritis	0.890 ± 0.071	Khoshvaghti *et al.* 2009
Subclinical Mastitis	0.570 ± 0.063	Khoshvaghti *et al.* 2009
Mild Mastitis	0.47 ± 0.45	Eckersall *et al.* 2001
Moderate Mastitis	0.74 ± 0.72	Eckersall *et al.* 2001
Chronic Inflammation	0.18-2.0	Takahashi *et al.* 2006

On the basis of the information in the Table 3, it can be inferred that only a moderate accuracy can be reached by measuring the serum concentrations of Hp in dairy cows (sensitivity: 50% to 79%, specificity: 54% to 87%, respectively) (Dubuc *et al.* 2010a; Huzzey *et al.* 2011).

The broad spectrum of stressors resulting in the acute phase response cause an unspecified increase in Hp 50-100 times basal levels (Kim *et al.* 2005). Combination of acute phase proteins (Peterson *et al.* 2004), body temperature, and clinical exam would improve diagnosis and causation of the acute phase response (Ceron *et al.* 2005). Huzzey *et al.* (2009) proposed a sensitivity scale of Hp to identify cows in need of treatment. A cut off point of >0.4 mg/mL of plasma Hp was recommended to identify cows that need further clinical examination to determine disease status. Plasma Hp threshold of >1.4 mg/ml gave necessity for antibiotic treatment. These findings hold promise for the use of acute phase proteins not only as a biomarker for disease, but also fertility in dairy cattle.

Factors affecting detection of Hp and SAA

- Patterns of Hp release vary between species and degree of inflammation.
- Whilst serum Hp concentrations increase in dairy cows with acute metritis, there is no significant deviation from normal values in chronic metritis cases (Skinner *et al.* 1991; Smith *et al.* 1998a).

- Test characteristic of elevated Hp concentrations as an indication of metritis might be confounded by several other factors such as an increase in the first week after calving due to higher pre-partum NEFA concentration, normal calving-related tissue damage, dystocia, retained fetal membranes, twinning, mastitits, fatty liver syndrome (Yoshino *et al.* 1992; Hirvonen *et al.* 1999; Humblet *et al.* 2006; Chan *et al.* 2010; Dubuc *et al.* 2010b; Silvestre *et al.* 2011), negative energy balance (Crawford *et al.* 2005), and clinical onset of disease in early lactation (Sabedra, 2012). Kovac *et al.* (2009) studied the relationship between acute phase proteins (Hp and SAA) and energy metabolites (NEFA and BHBA) in dairy cows and found significant correlations between Hp with NEFA as well as BHBA and SAA with only NEFA. Hence, Hp as an indicator of infection should be used with caution in the days following parturition (Humblet *et al.* 2006). Interestingly, increased Hp concentration during the first week postpartum was found as a common risk factor for all these conditions. The findings suggested that despite variation in clinical manifestations of postpartum uterine infections, Hp could serve as a better indicator for the postpartum performance of dairy animals.
- There is a positive correlation between hormonal status (estrogens, progesterone, and cortisol) and Hp level during the last trimester of pregnancy and after calving (Young *et al.* 1995; Alsemgeest *et al.* 1996).
- Horadagoda *et al.* (1999) reported that assays of Hp and SAA might differentiate the chronic from the acute inflammation in cattle in a better way than the hematological tests. In contrast, Humblet *et al.* (2006) suggested that Hp and SAA were good markers for identification of healthy animals, but their ability to identify the diseased animals was low as most of the postpartum diseases were chronic in nature. Further, they reported that physiological status of a cow, particularly at parturition, can have an influence on serum Hp concentration.
- Hp is not influenced by environment or physiological status and thus a useful indicator for cows with postpartum reproductive disorders (Chan *et al.* 2010).
- Serial Hp measurements have also been utilized to evaluate the efficacy of different therapeutic regimens. Smith *et al.* (1998b) evaluated the therapeutic efficacy of different antimicrobial regimens through serum Hp concentrations in cows that suffered with toxic puerperal metritis and found that the 5-day treatments reduced the serum Hp concentration (19 mg/dL *vs.* 7.35 mg/dL). Heidarpour *et al.* (2012) reported that Hp level was reduced after treatment in cows that suffered with clinical endometritis and cows had a lower concentration of Hp (before treatment), showed a better response to further

treatment. In contrast, Jeremejeva *et al.* (2012) evaluated the effect of two treatment strategies (ceftiofur, NSAID, ceftiofur, PGF2a) in acute puerperal metritis and clinical metritis cows through SAA and Hp and found no significant difference. Mordak (2008) found that the cows with highest Hp concentration expelled placenta after 4 days (2.22 g/L) and cows with the lowest concentration of Hp (0.9 g/L) had an easy placental removal.

- It has been shown that feeding certain types of grains like barley stimulate an inflammatory response. The use of feeds to alter the pH balance of the rumen for a short time could potentially increase acute phase proteins in circulation to overcome an immune challenge. An increase in SAA and lipopolysaccharide-binding protein that have shorter half-lives than Hp and bind endotoxins in circulation drawing leukocytes towards the complex (Emmanuel *et al.* 2008).
- Primiparous animals exhibit a higher acute phase reaction when compared to multiparous animals (Cullens, 2005; Saremi *et al.* 2012).

In view of the findings that the illness status was associated with plasma Hp and milk yield (low<41.95 kg, high>41.96 kg) with higher plasma Hp concentrations having higher illness rates at 54 to 60 days in milk, but not at 20 to 26 days in milk. It was concluded that the combination of calving difficulty, clinical signs of illness, uterine culture, and Hp plasma concentration (after 26 days in milk) produced in the most predictive model of fertility to the first AI around day 60 postpartum and therefore need to be considered for a genuine interpretation (Wise, 2016).

Procalcitonin (PCT) and others

A standard blood examination in presence of sepsis, predominantly in human medicine, a search for potent biomarkers to differentiate sepsis from other inflammatory processes is ongoing. Current advice in human medicine is to initiate broad spectrum antimicrobial therapy before culture results are returned in patients above this cut-off. Acute phase proteins are known to be highly species specific. In calves, PGE2, malondialdehyde (MDA), IL-8, TNF-α, IFN-y, neopterin and PCT have been evaluated in a single study (Ercan *et al.* 2016). The last four were significantly increased in *E. coli* septicemia calves. In addition, C - reactive protein (CRP) has also been found to elevate in bacterial infections (Simon *et al.* 2004). It has been suggested that the measurement of PCT, neopterin, acute phase proteins, and cytokines together with the clinical examination may be a useful diagnostic and prognostic tool for assessment of calves naturally infected with *M. haemolytica* and *H. somni* (El-Deeb *et al.* 2020). Especially PCT, has been put forward as a suitable sepsis marker in infants, with the optimal cut-off being >1.2 ng/mL (sensitivity = 77%; specificity = 79%) (Delevaux *et al.* 2003; Park *et al.* 2014). It has also been stated that PCT levels above 0.5 ng/mL may be an indicator of diseases in human medicine (Luyt *et al.* 2005).

PCT is a prohormone of calcitonin and is a 116 amino acid protein with a molecular mass of 13 kDa (Lee, 2013). In a normal metabolic state, hormonal active calcitonin is produced and secreted by C-cells of the thyroid gland after proteolytic treatment processing of PCT intracellularly. Under normal conditions, negligible serum PCT concentrations are detected (Whicher *et al.* 2001). It has been reported that in infective cases modulated by lipopolysaccharides and sepsis-related cytokines, PCT is secreted from the extrathyroidal source such as liver, lung and peripheral blood mononuclear cells (Oberhoffer *et al.* 1999; Meisner, 2002; Lee, 2013; Dong *et al.* 2019). Serum PCT level is associated with the prevalence and severity of bacterial infections. It has also been reported that increased interferon gamma (INF-Ò) during viral infections suppresses PCT production and may be useful in differentiating viral and bacterial infections (Briel *et al.* 2008; Schuetz *et al.* 2011). The kinetics of a prospective marker should be considered along with its sensitivity and specificity. PCT started to increase 4 h after inflammation, peaked about 6 -8 h later (Dandona *et al.* 1994; Brunkhorst *et al.* 1998; Nijsten *et al.* 2000), and rapidly returned to normal values after the inflammation was controlled by the organism (Dandona *et al.* 1994; Meisner *et al.* 2001). PCT is stable in samples, the assay is relatively easy to perform, with a moderate cost and the result is available within 2 h (Enguix *et al.* 2001). The latter findings were echoed by Ercan *et al.* (2016) who also suggested PCT to be the best marker as it increased four times, had a long half-time life and remained stable at room temperature. In order to be practically useful for on-farm decision-making, the development of a cow-side test with this molecule will be necessary.

SAA and PCT were evaluated in certain inflammatory diseases of cattle including the cows without systemic clinical symptoms, but with a purulent or mucopurulent discharge from the uterus (clinical endometritis) 21 days or more after birth. The cut-off value of SAA and PCT for healthy and inflammatory cattle was determined 28.52 µg/mL and 149.55 pg/mL (Basbug *et al.* 2020). PCT levels may be used as an alternative to serum SAA measurement as an indicator of acute inflammation in cattle. The findings were based on the reason of finding serum PCT concentrations to be ~9 times higher in the infected cattle, especially those with peritonitis than in the healthy cattle. Besides being infection specific marker, the other advantage of relying on PCT would be that viral diseases and autoimmune diseases do not cause increases in PCT. Therefore, in human medicine PCT is most commonly used to distinguish between bacterial diseases and nonbacterial diseases (Alkholi *et al.* 2011; Joo *et al.* 2011). In addition, serum PCT levels were found to be high in sepsis, bacteremia, meningitis and fungal infections causing serious systemic infection (Castelli *et al.* 2004; Dornbusch *et al.* 2005; Alkholi *et al.* 2011). This observation is consistent with the findings of previous researches and may be explained by the fact that IFN-Ò inhibits IL-1 beta-induced calcitonin mRNA expression and PCT secretion (Briel *et al.* 2008; Schuetz *et al.* 2011; Ercan *et al.* 2016). All values of PCT above 0.5 ng/mL are considered important and indicate that patients are in life threatening conditions (Meisner, 2002).

C – Reactive protein: It is an acute-phase reactant, and CRP level measurements are frequently used to aid in the diagnosis of bacterial infections. CRP is synthesized by the liver, mainly in response to IL-6, which is produced not only during infection but also in many types of inflammation (Ridker, 2003). It binds to polysaccharides in pathogens, activating the classical complement pathway. The reported diagnostic accuracy of PCT and CRP for the diagnosis of bacterial infections has varied across studies. CRP secretion starts within 4 to 6 h after stimulation, peaking only after 36 h. The assay to determine CRP levels is easy to perform, often automated, and has a low cost (Enguix *et al.* 2001).

It was observed that the PCT measurements were performed using the same commercially available specific antibody system. However, means of measuring CRP levels largely varied, with 8 different methods used among 12 different studies. The implications of multiple assay methods are unknown in the final result of meta-analysis. However, each study included used its own best cutoff value, and the linear regression methods used in the analysis accounted for possible threshold differences between studies (Simon *et al.* 2004).

o. Hysteroscopy

A non-invasive approach for the diagnosis of clinical endometritis in cows could be hysteroscopy (Madoz *et al.* 2010). However, due to the high cost of the equipment and the problems encountered in passing the bovine cervix, the prospects of such a technique are currently limited and suggested for specialized cases only (Milosh *et al.* 2013).

p. Electronic Nose

In human medicine, electronic nose devices have been used for different purposes including diagnostics of urinary tract infection (Pavlou *et al.* 2002), wound infection (Setkus *et al.* 2006), and tuberculosis (Bruins *et al.* 2013). In bovine research, electronic nose devices were used for the detection of estrus, mastitis, or respiratory disease (Eriksson *et al.* 2005; Knobloch *et al.* 2010; Wiegerinck *et al.* 2011). Data on sensitivity and specificity, however, were not described.

Principle: The use of electronic nose in reproduction is based on the principle that it can differentiate between the normal and abnormal samples on the basis of the amount of volatile organic compounds of C2-C4 alcohols, linear and cyclic ketones, aldehydes, aliphatic amines, and sulfur-containing organic compounds, acids and ammonia.

The rationale of using an electronic nose device (Figure 11) for uterine infections has been quoted with a threefold advantage (Burfeind *et al.* 2014a)

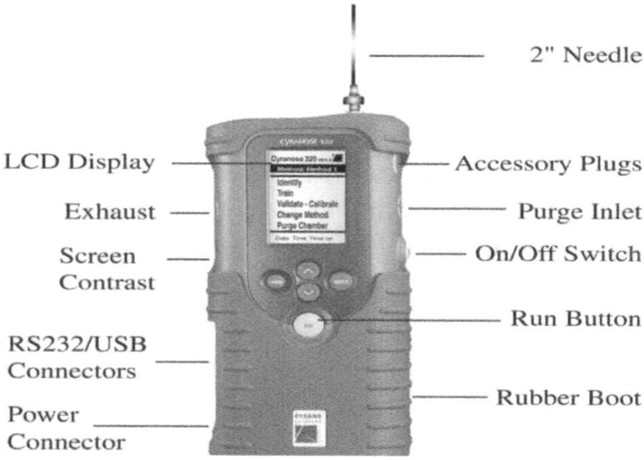

Figure 11: Electronic nose (DiagNose, C-it, Zutphen, the Netherlands)

i. **Eliminates the interpretation bias between observers**: The intra-assay repeatability of odor assessment of vaginal discharge by means of an electronic nose device was higher compared with the human nose (Sannmann *et al.* 2013a). It has been demonstrated that sensorial assessments such as the olfactory evaluation of odor of vaginal discharge or the visual evaluation of vaginal fluid through vaginoscopy can be confounded by only moderate intra- and interobserver repeatability (Leutert *et al.* 2012). For instance, a moderate interobserver (K=0.43) and intraobserver repeatability (K=0.52) for the classification of healthy *versus* metritic animals based on the assessment of vaginal discharge by olfactory cognition (Sannmann *et al.* 2013a). Hence, the element of individual variation can be negated using electronic nose.

ii. **Identification of the infecting microbes:** The composition of the bacterial flora in the postpartum uterus differs between healthy and metritic cows (Santos *et al.* 2011). Previous research in human medicine had demonstrated that electronic nose devices can be used for identification of bacterial pathogens such as *E. coli*, *Proteus* spp., and *Staphylococcus* spp. (Pavlou *et al.* 2002; Bruins *et al.* 2009). The findings of Burfeind *et al.* (2014a) demonstrated that acute puerperal metritis can be detected accurately by the Diag-Nose, although it is not possible to determine the causing pathogen. It was assumed that the opportunistic nature of the acute puerperal metritis infection makes is difficult to determine the causing pathogen especially in mixed cultures, which made up a large proportion of the bacteriologic results. Hence, insptie of change in odor due to infection, the change might not be directly related to the type of pathogen.

iii. **Identification of the bacterial load**: Electronic nose is capable of detecting different volatile organic compounds commonly produced and released from organic sources such as living microbes and multicellular organisms (Barsan *et al.* 2007; Bruins *et al.* 2013; Wilson and Baietto, 2011). Odor of vaginal discharge at 21 and 28 days postpartum is associated with the bacterial growth density of potential pathogens (i.e., *Escherichia coli*, *Trueperella pyogenes* and *Fusobacterium necrophorum* in the uterus (Williams *et al.* 2005). Therefore, abnormal vaginal discharge is a plausible criterion, which is commonly used in the field because it is easy and intuitively assessable without the use of additional diagnostic tools.

In a study (Skorikov *et al.* 2021), samples of uterine secretions were obtained on days 1,3,6,8 after parturition. A MAG-8 device (Russia) was utilized for the purpose having eight sensors of different piezoelectric crystals frequency of 10.0 MHz and film sorbents on electrodes that responded to the different compounds of C2-C4 alcohols, linear and cyclic ketones, aldehydes, aliphatic amines, and sulfur-containing organic compounds, acids, ammonia. During the first three days after calving in cows with postpartum endometritis in comparison with healthy animals, the area of the visual response of the sensors revealed an increase of 28.4% of the alcoholic content, linear structure ketones by 5.2-6.0%, whereas the aliphatic acids and alkyl amines were reduced by 14.3% and 9.2%, respectively. On 6-8 days after calving, in the clinical endometritis cows, an excess of the "visual imprint" by 6.1 times, a decrease in the content of aliphatic acids by 22.7%, and an increase in the concentration of sulfur-containing organic compounds by 4.9%, was observed. On the basis of the findings it was inferred that the electronic nose system can distinguish between the normal and pathological composition of the volatile compounds in the postpartum cervical secretions.

Another study (Sannmann *et al.* 2013a) using an electronic nose (DiagNose, C-it, Zutphen, the Netherlands) reaffirmed that the accuracy for evaluating vaginal discharge odor can be increased using the device, however, the system needs further development for the field use. Burfiend *et al.* (2014a) found that using a threshold value of 0.3, the best observed sensitivity and specificity of electronic nose in acute puerperal metritis was 100% and 91.6%, respectively.

q. Intrauterine Oxygen Reductase Potential (Eh)

Eh may be used in assessing the degree of bacterial contamination of the uterine lumen. Eh values fell in the presence of infection (El-Azab *et al.* 1988), thereby creating an anaerobic environment; this drop may be associated with either micro-organism metabolism or increased oxygen consumption by PMN cells.

D. Comparison of different diagnostic techniques

a. Transrectal Palpation *Versus* Culture

For clinical endometritis transrectal palpation is most common means of diagnosis, but compared to uterine culture palpation correctly identified only 22% of cows predicted to have uterine infection (Miller *et al.* 1980).

b. Ultrasound (uterine fluid) *Versus* Cytology

Endometritis has been diagnosed by detection of fluid in the uterus using ultrasonography (Kasimanickam *et al.* 2004). Nonetheless, this method was found to be less sensitive than endometrial cytology (Barlund *et al.* 2008). Similarly, the diagnostic methods, such as ultrasonographic evaluation of the reproductive tract and uterine content are inferior to cytologic examination of uterine content (Kasimanickam *et al.* 2004; Barlund *et al.* 2008), which lead to the proposed disease definition based on cytology as the presence of >18% neutrophils in the uterine samples collected between 21 and 33 days postpartum, or>10% neutrophils between 34 and 47 days postpartum, in the absence of purulent vaginal discharge (Sheldon *et al.* 2006). In a farm setting, however, the latter method is inconvenient, as it involves collection of the sample, preparation of the slides and staining, followed by microscopic examination and identification and enumeration of cells. Uterine samples collected using the cytobrush method (Barlund *et al.* 2008; Kasimanickam *et al.* 2005b) allow easier slide preparation compared with samples collected using low-volume uterine lavage, but still require the time-consuming cell evaluation step.

c. Cytotape *Versus* Cytobrush

Agreement in terms of the percentage of PMNs in both the methods was good with a minor standard error of 2%. Both methods yielded similar total cellularity (P ¼ 0.62). Cytotape yielded better quality smears with more intact cells while samples that were taken by cytobrush were more likely to be bloody (Figure 12). Hence, cytotape and cytobrush methods yielded smears with a similar PMNs percentage and a total number of cells, but cytotape provided smears with higher quality and significantly less blood contamination. In another experiment, cytology slides were stained using both Diff-Quik or CIAE (specific staining method for PMNs), naphthol-AS-D-chloroacetate-esterase. Because the CIAE staining method requires several preparatory steps including a time-consuming incubation, it is considered time demanding and relatively expensive. Therefore, this technique is not recommended for routine use under field circumstances. Agreement between PMNs percentage in both staining techniques was good with a standard error of only 2%. Hence, Diff-Quik was confirmed as an easy, fast, and high-quality staining technique, which can be routinely used to stain endometrial cytology samples satisfactorily. Consequently, cytotape offers similar characteristics as the cytobrush, but requires less specialized material and can be used during AI (Pascottini *et al.* 2015).

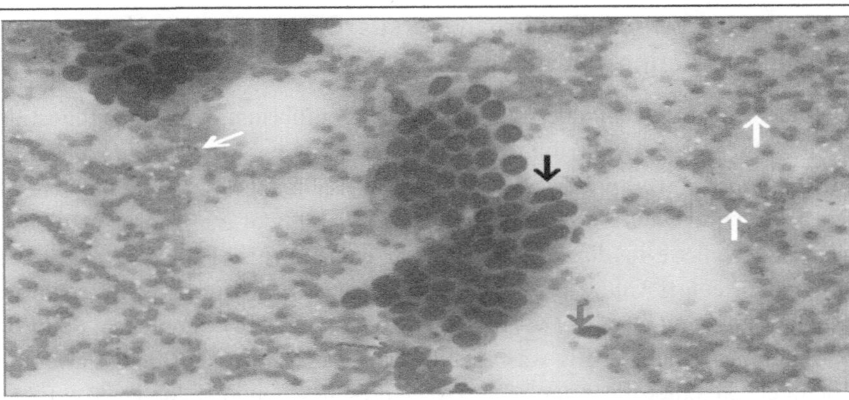

Figure 12: An uterine cytobrush smear depicting abundance of RBCs (Red arrows = PMN cells; White arrows = RBCs; Black arrows = Endometrial cells)

d. Cytobrush *Versus* Low Volume Uterine Lavage

Both, cytobrush and low volume lavage are acceptable cytology techniques to diagnose subclinical endometritis in dairy cows, but the principal reason why cytobrush is preferred by practitioners is due to its practicability and cytologic quality (Kasimanickam *et al.* 2005b; Barlund *et al.* 2008). The main advantage of low volume lavage is that samples are taken from a larger endometrial surface and tend to be more representative for the health of the complete endometrium (Miller *et al.* 1980; Roszel and Freeman, 1988; Bonnett *et al.* 1991b; Bourke *et al.* 1997; Bohn *et al.* 2014). Major disadvantage of low volume lavage is related to the difficulty to recover fluid after the infusion (17% of the cases), damage due to manipulation during sampling, irritation that flushing liquid can cause to the endometrium and a higher percentage of distorted cells, in comparison to cytobrush smears (Ball *et al.* 1988; Roszel and Freeman, 1988; Brook, 1993; Kasimanickam *et al.* 2005b; Barlund *et al.* 2008; Overbeck *et al.* 2011; Cocchia *et al.* 2012; Bohn *et al.* 2014; Dini *et al.* 2015).

Elaboration of the findings of Kasimanickam *et al.* (2005b) who compared the two sampling techniques concluded that cytobrush had some advantages over uterine lavage: it was less time-consuming, was easier to perform independently of the uterine size, did not produce endometrial irritation, and induced lower degree of cell structure distortion and lower presence of erythrocytes. One disadvantage of cytobrush is that the sample is collected from a specific area of the endometrium, whereas uterine lavage provides cells from the whole endometrial surface.

e. Cytotape *Versus* Cytobrush *Versus* Cotton Swabs *Versus* Low Volume Uterine Lavage

The ideal cytology technique should represent equilibrium between practicability and the possibility to yield highly accurate results. A harmless technique yielding a high number of well-preserved cells is indispensable for reliable cytologic results (Roszel and Freeman, 1988). Considering the reliability of endometrial cytology as the best technique to diagnose subclinical endometritis (Dubuc *et al.* 2010a), variation in cutoff values to differentiate affected *versus* unaffected cows was the main problem (Madoz *et al.* 2013; De Boer *et al.* 2014). A practical definition for subclinical endometritis was established by consensus in 2006, stating that subclinical endometritis is considered to be present when in endometrial cytology, samples taken between 21 and 33 days postpartum greater than 18% of harvested cells are identified as PMNs, or greater than 10% of the cells are PMNs when samples are taken at 34 to 47 days postpartum, in the absence of clinical endometritis (Sheldon *et al.* 2006).

In human medicine, the phenomenon of cellular distortion and fragmentation was described when the cotton swab was used (Cocchia *et al.* 2012; Bohn *et al.* 2014). This distortion and fragmentation are mainly related to the firm adherence of cells to the cotton fibers and the consequent pressure needed to take the samples and to roll them onto the glass slide (Bourke *et al.* 1997; Trimbos and Arentz, 1985). These limitations were overcome to a large extent by using cytobrush, which yielded significantly more cells than low volume lavage or cotton swab (Trimbos and Arentz, 1985; Overbeck *et al.* 2011; Cocchia *et al.* 2012). Hence, cytobrush was recommended as the technique of choice because of its feasibility, safety, and reasonably high-quality samples (Kasimanickam *et al.* 2005b; Barlund *et al.* 2008).

Cytology samples to diagnose subclinical endometritis were obtained by cytobrush (Kasimanickam *et al.* 2004) or low-volume uterine lavage (Gilbert *et al.* 2005), either of them providing similar results (Kasimanickam *et al.* 2005b; Barlund *et al.* 2008). However, there seems not to be an ideal cytology technique; each of the three methods has specific advantages and limitations (Cocchia *et al.* 2012). In low volume lavage the cells can be diluted (Kasimanickam *et al.* 2005b). Although, cytobrush slides yielded a better quality in comparison to the cotton swab slides due to the rigidity of the brush bristles, the fragmentation of cells when the brush is rolled on the microscope slide is a limitation of the cytobrush (Cocchia *et al.* 2012).

Cytotape was found to be better than cytobrush. In terms of cellularity, both cytotape and cytobrush yielded a similar cellularity, but by virtue of paper tape in cytotaping device, it does not have absorbent properties as the cotton swab, and most of the cellular material collected was adhered to the glass slide when the tape was rolled on it. In terms of cellularity, both quality and RBC contamination were found to be significantly better for cytotape in comparison to cytobrush. More intact cells were found in cytotape slides, resulting in significantly more cytotape samples being categorized in the category "very good".

Another limitation of cytobrush was that it posed a significantly higher risk to have a bloody smear in comparison to cytotape sampling. Presence of higher numbers of RBCs in endometrial cytology after low volume lavage compared to cytobrush was observed (Kasimanickam *et al.* 2005b) in cows, and more bloody samples in cytobrush *versus* cotton swab (Bourke *et al.* 1997) in mares. A robust difference was represented by around 35% of cytobrush samples with a high or moderate contamination of RBCs. Rigid fibers of the cytobrush might be responsible for high amount of RBCs in the samples (Cocchia *et al.* 2012), and this blood contamination should not be regarded as a negligible observation. Regarding the high amount of bloody samples when using the cytobrush, one might argue damaging the uterine epithelium and the concomitant production of lesions in the endometrium, which opportunistic bacteria might use as an entrance gateway to cause a secondary infection. Additionally, bloody contamination of the samples implies the presence of 1% or 2% of PMNs (Kahn *et al.* 2005). Hence, when a low threshold of PMNs is used to diagnose subclinical endometritis (Salasel *et al.* 2010; Pothmann *et al.* 2015), these PMNs by virtue of endometrial damage could interfere with the diagnosis.

f. Ultrasonograpy *Versus* Histopathology

In cows that were 21 to 47 d postpartum, without vaginal catarrh, an ultrasonographic and a uterine biopsy were preformed. A combination of the determination of intrauterine fluid and cervical diameter through ultrasound was the most efficient ultrasonographic method, with 50% sensitivity and 88% specificity. Histopathological examination, on the other hand, provided rich detail on endometrial inflammation, and allowed evaluation of different components of the tissue such as the epithelium, lamina propria, endometrial glands, and blood vessels; this method had 44% sensitivity and 92% specificity for diagnosis of endometritis (Meira Jr. *et al.* 2012).

g. Cervix *Versus* Intraluminal Fluid (ultrasonography)

LeBlanc *et al.* (2002a) showed that animals with a delayed cervical involution often demonstrate poorer reproductive performance compared with animals diagnosed as normal. Although being a subjective assessment, determination of fluid in the uterus was a reliable technique, having good specificity, as documented in other studies (Mateus *et al.* 2002a; Kasimanickam *et al.* 2004; Barlund *et al.* 2008; Oral *et al.* 2009; Meira Jr. *et al.* 2012). It was found that the efficiency of cervical measurement for the diagnosis of endometritis can be increased when combined with the presence of intraluminal fluid and constitutes the best ultrasonographic examination to assess uterine postpartum health (Meira Jr. *et al.* 2012).

h. Pus *Versus* Cervical Diameter

Clinical endometritis has been diagnosed by the presence of purulent uterine discharge or cervical diameter >7.5cm, or mucopurulent discharge in cows after 26 days in milk (LeBlanc, 2008). Similarly, the presence of purulent leaks and a cervical diameter >7.5 cm at 20 days postpartum, or mucopurulent discharge at 26 days are signs of clinical endometritis (LeBlanc *et al.* 2002 a), which have been used to diagnose clinical endometritis.

i. Vaginoscopy *Versus* Transrectal Palpation for Detecting Uterine Discharge

Miller *et al.* (1980) concluded that vaginoscopic examination is a more accurate method for detecting uterine infections than palpation transrectal. In the study by LeBlanc *et al.* (2002a), the reproductive performance of cows with purulent discharge on vaginoscopic examination was significantly lower than that of cows with no abnormal discharge. However, vaginoscopy often failed to identify all cows that are truly at risk of poor reproductive performance. Hence, the absence of uterine discharge by vaginoscopic examination was not truly indicative of absence of uterine inflammation. The vaginoscopic identification of discharge may be influenced by the severity of the infection, myometrial contraction, uterine clearance mechanisms, perineal conformation, body condition score, postural changes, and exercise. Increased production of cervical mucus and myometrial contractions during estrus may increase the probability of observing discharge externally or in the anterior vagina. Alternatively, uterine discharge may not be detected externally or by vaginoscopy in cows in which the cervix is closed so that the cows that may have abnormal uterine fluid but fail to show clinical signs, lead to false negative results. Similarly, cows with subclinical endometritis may not be identified by routine diagnostic procedures.

j. Endometrial Biopsy *Versus* Bacterial Culture

Lymphocytic nodules and inflammation in the endometrium, for example, have been assumed to be associated with bacterial infection, although many "infected" animals have not shown abnormalities of the endometrium, and not all cows with reproductive abnormalities have had bacteria isolated concurrently from the uterus (Sagartz and Hardenbrook, 1971; Studer and Morrow, 1978). However, the histopathology of uterine biopsy and bacteriogical culture of uterine content have been considered the gold standard for diagnosis of endometritis (Gilbert *et al.* 2005; Sheldon *et al.* 2006).

Actinomyces pyogenes were the most common isolates in postpartum cows. Isolation of *A. pyogenes* from a biopsy at day 26 postpartum was positively correlated with isolation of anaerobic bacteria and segmented cell inflammation in the same biopsy, and with subsequent isolation of *A. pyogenes* at day 40. There was a strong association between isolation of *A. pyogenes* and anaerobes at day 26 with increased uterine lesions at day 40. Isolation of alpha hemolytic streptococci was negatively correlated with isolation of *A. pyogenes* and with inflammation. *A. pyogenes* and alpha hemolytic streptococci showed opposite associations with mononuclear cell inflammation and lymphocytic foci (Bonnett *et al.* 1991a).

k. Biopsy *Versus* Bacterial Culture *Versus* Whiteside Test

Twenty one repeat breeder cows were positive for Whiteside test and 12 showed growth in bacteriological culture with different degrees of changes in the endometrium. It was concluded that histopathology of the endometrium may clearly indicate the future fertility status when compared to Whiteside test and bacteriological culture analysis. Twenty per cent cows in control showed mild form of infection based on Whiteside test and ten percent showed growth of aerobic organism. The histopathology of endometrium showed marked cellular changes of different degree, such as glandular hyperplasia (13/21), moderate glandular hyperplasia (6/21), degenerative changes of the endometrium (13/21), cystic dilatation (2/21), mononuclear cell infiltration (3/21) and perivascular fibrosis (1/21). Among the repeat breeding cows which had periglandular fibrosis, 52.3% (11/21) showed involvement of 5 to 10 layers, 23.8% (5/21) 10 to 15 layers, and 9.5% (2/21) >15 layers. Twenty percent metrorrhagia and 40 percent PMN cell infiltration were observed in control animals (Raja *et al.* 2012).

l. Cyobrush *Versus* Vaginoscopy *Versus* Ultrasound

No significant difference was observed between vaginoscopy and ultrasound when compared to cytobrush technique. Hence, endometrial cytology could be used in cows safely and effectively for the diagnosis of infertility particularly in cases of subclinical endometritis. Furthermore, it was suggested that cytobrush technique could also be used during the follow up of the recovery period (Oral *et al.* 2009).

m. Leucocyte Esterase *Versus* Albumin *Versus* Blood Urea Nitrogen *Versus* Myeloperoxidase

Plasma proteins, including albumin, act as carriers for the transport of substances, such as fatty acids, during inflammation (Mazzaferro *et al.* 2002), and blood urea nitrogen (BUN) levels reflect protein metabolism in dairy cows (Butler, 1998). Previous studies have reported that there is a relationship between uterine diseases and concentrations of albumin, BUN, and total cholesterol (Kaneene *et al.* 1997; Van Sun, 2004; Ghanem *et al.* 2016). Most dairy cows experience a negative energy balance status close to and after calving, and the concentrations of NEFA and ketone bodies increase substantially after calving (Bertics *et al.* 1992; Valergakis *et al.* 2011).

Leucocyte esterase level had a significant association with PMN% at all weeks postpartum. Albumin and BUN levels had weak negative associations with endometrial PMN% at week 3. Pathogenic bacterial detection rates were higher in the cows with endometritis at week 3 and week 5. Conception rate at first AI (mostly at week 7) tended to be lower in the cows diagnosed with endometritis at week 3 than in the healthy cows. In conclusion, associations were found between endometrial leucocyte esterase and endometritis, but not for myeloperoxidase and endometritis. Diagnosing endometritis at week 3 may be the best moment to predict subsequent fertility. Both myeloperoxidase and leucocyte esterase were detected in the cytobrush samples instead of low volume lavage. Use of ELISA procedure for myeloperoxidase estimations makes it an expensive procedure.

n. Leukocyte Esterase *Versus* Endometrial Cytology *Versus* Vaginal Discharge Score

Accuracy of leucocyte esterase test, endometrial cytology and vaginal discharge score for diagnosing postpartum reproductive tract health status in dairy cows at the moment of sampling has been evaluated in different studies. In the initial studies, the thresholds used in dairy practice comprised of leucocyte esterase ≥ 1 (small amount of leukocytes); endometrial cytology of 6% of PMN and vaginal discharge score of ≥ 3 (mucopurulent discharge). Leucocyte esterase had the highest sensitivity (65.6%) and the lowest specificity (77.9%) among the three tests (Arango-Sabogal *et al.* 2019). A similar sensitivity of endometrial cytology and vaginal discharge score was observed (around 50%), with the vaginal discharge score having the highest specificity (99.6%) (Dubuc *et al.* 2010a; Denis-Robichaud and Dubuc, 2015). Denis-Robichaud and Dubuc (2015) concluded that a combination of either leucocyte esterase of ≥ 1 (small amount of leukocytes) or endometrial cytology of ≥ 6% of PMN with vaginal discharge score of ≥ 4 (purulent discharge or worse) gave the optimal ability to predict subsequent nonpregnancy at first service. Dubuc *et al.* (2010a) suggested that the optimal vaginal discharge threshold for diagnosing endometritis was a score of ≥ 3 (mucopurulent discharge or worse) when performed either at 35±3 d or 56±3 d postpartum. In a subsequent study, a different statistical model was used, whereby the optimal thresholds were: leucocyte esterase ≥ 2 (moderate amount of leukocytes), ≥ 6% neutrophils on endometrial cytology, and vaginal discharge score of ≥ 2 (mucus with flecks of pus). The vaginal discharge score had the highest sensitivity and specificity among the tests evaluated. The leucocyte esterase at a threshold ≥ 2 (moderate amount of leukocytes) had sensitivity and specificity comparable to that of endometrial cytology.

o. Vaginoscopy *versus* cytology

Barlund *et al.* (2008) reported that vaginoscopy lacked sensitivity when compared to endometrial cytobrush cytology for the diagnosis of both clinical and subclinical endometritis, but did support the use of vaginoscopy for the diagnosis of clinical endometritis in cows > 4 weeks postpartum.

Furthermore, Dubuc *et al.* (2010a) found a poor agreement between cytological endometritis diagnosed by uterine cytology and vaginal discharge. Although the term endometritis is well established for cows with vaginal discharge, there is clear evidence that the origin of pus in the vagina is not always the endometrium but could also be the cervix or vagina.

p. Ultrasound *Versus* Biopsy *Versus* Cytology *Versus* Culture

Ultrasonography (Barlund *et al.* 2008), culture of bacteria from uterine fluid (Bretzlaff, 1987), uterine biopsy (Bonnett *et al.* 1991a), and uterine cytology, have been described for diagnosis of endometrial infection or inflammation (Gilbert *et al.* 2005). Ultrasonography and culture of bacteria from uterine fluid have been reported to have low sensitivity, and uterine biopsy followed by histological examination is expensive, time-consuming, and clinically inaccessible under practical field conditions. However, endometrial cytology using cytobrush techniques to know the PMN% appears to be of considerable importance to be able to correctly diagnose postpartum endometritis using validated cow-side techniques to improve the reproductive performance of dairy cows (Baranski *et al.* 2012).

q. Myeloperoxidase *Versus* Cytology

High myeloperoxidase levels in milk has been reported to be related to high somatic cell numbers and neutrophil count in cows (Cooray, 1994). Although mares with positive cytology had significantly high myeloperoxidase levels (Parrilla-Hernandez *et al.* 2014), such an association was, however, missing in the dairy cows (Nazhat *et al.* 2018). One possible explanation for the discrepancy between the findings between mares and cows may emerge from the estrous cycle perspective. While endometrial samples from mares were evaluated during estrus, the samples in cows were collected during the early postpartum period regardless of the estrous cycle. Moreover, studies clarifying myeloperoxidase profiles during the estrous cycle in cows are required to confirm whether myeloperoxidase levels change with the stage of the estrous cycle.

r. Ultrasound (intrauterine fluid and uterine diameter) *Versus* Cytology

Ultrasonography and uterine cytology on swabs or fluids collected from uterus have been considered accurate for the diagnosis of subclinical endometritis. It was felt that uterine cytology to diagnose subclinical endometritis is not practical under most clinical settings (Madoz *et al.* 2014). Therefore ultrasound examination may be used to identify some, but not all cases of subclinical endometritis (Kasimanickam *et al.* 2004). A small amount of fluid in the uterine lumen and/or thickened uterine walls, as assessed by ultrasound, were considered as signs of endometrial inflammation. However, in various studies ultrasound was found to be less sensitive than endometrial cytology (Kasimanickam *et al.* 2004; Barlund *et al.* 2008; Meira Jr. *et al.* 2012) for subclinical endometritis diagnoses. Presence of intrauterine fluid and a thick uterine mucosa may be normal findings in physiological situations such as estrus or early postpartum (Quintela *et al.* 2016), and perhaps the evaluation of fluid characteristics could improve the sensitivity of ultrasound diagnosis (Lopez-Helguera *et al.* 2012). Fuentes *et al.* (2017) found a significant relationship between presence of abnormal intrauterine fluid and subclinical endometritis diagnosed by biopsy, but not by cytology.

Debertolis *et al.* (2015) found significantly increased blood flow in uterine arteries of cows to which acute endometritis had been experimentally induced. Whether patterns of vascular flow may differ between healthy uterus and those with subclinical endometritis still has to be investigated. The latter study, performed transrectal Doppler sonography on uterine arteries of six cyclic cows before and for 4 days after inducing acute endometritis by intrauterine infusion of 720 mg of policresulen, and for 4 days of the following estrous cycle. Time-averaged maximum velocity (TAMV) increased and pulsatility index (PI) decreased significantly within 1 h of policresulen administration, and did not change in the next 4 days of the same cycle. TAMV and PI values in the subsequent cycle did not differ from the values measured before infusion and showed no changes within the cycle. All cows exhibited an acute endometritis determined by histopathological findings of biopsy samples taken 1 day after infusion and fibrotic endometrial alterations detected in the subsequent cycle. No relationships were observed between fibrotic changes of the endometrium and uterine blood flow during either cycle. In conclusion, acute inflammation is accompanied by a rise in uterine blood flow, but fibrotic alterations do not seem to be related to Doppler sonographic findings.

s. Ultrasound (fluid volume) *Versus* Bacterial Growth

Mateus *et al.* (2002a) examined postpartum cows by transrectal ultrasonography and concluded that the volume of intrauterine fluid was significantly associated with impaired uterine involution and the intrauterine fluid volume score was positively correlated with bacterial growth.

t. Ultrasound (endometrial thickening) *Versus* Cytobrush

It seems reasonable that a local inflammatory response within the endometrium would result in some degree of tissue thickening. Barlund *et al.* (2008) reported that endometrial thickness (>8mm) measurements were less useful than endometrial cytobrush cytology to diagnose subclinical endometritis between 28 and 41 days postpartum.

u. Cytobrush *Versus* Biopsy

Several studies (Madoz *et al.* 2013; Pascottini 2016; Fuentes *et al.* 2017) have found that the PMN% in cytological samples taken with cytobrush was not affected by the stage of the estrous cycle. However, when the subclinical endometritis diagnosis was done by biopsy, a higher degree of inflammatory infiltration could be observed in the follicular phase of the cycle (Fuentes *et al.* 2017). This was because cytology only detects PMN infiltration in the superficial epithelium, whereas biopsy allows identifying inflammatory cells in deeper layers of the endometrium. Fuentes *et al.* (2017) compared cytology and biopsy findings between the two horns of 100 bovine uteri collected postmortem and observed that samples collected from the left horn were more representative of both uterine horns than those collected from the right one.

v. Transrectal (cervix, uterus, ovaries) *Versus* External Inspection *Versus* Vaginoscopy

Le Blanc *et al.* (2002a) examined cows once between 20 and 33 days in milk including external inspection, vaginoscopy, and transrectal palpation of the cervix, uterus, and ovaries. The presence of purulent uterine discharge or cervical diameter >7.5 cm after 20 days in milk, or mucopurulent discharge after 26 days in milk identified cows with clinical endometritis. Given vaginoscopy, no diagnostic criteria based on palpation of the uterus had predictive value for time to pregnancy. Vaginoscopy identified 44% of such cases.

w. Leucocyte Esterase *Versus* pH *Versus* Protein

Cheong *et al.* (2012) evaluated the use of leucocyte esterase, protein, and pH tests using widely available urinary test strips (Multistix 10 SG; Bayer Corporation,

Elkart, IN, USA) on uterine lavage samples for the diagnosis of cytologic endometritis (Figure 10). Protein and pH results were evaluated after 1 min and the leucocyte esterase result evaluated after 2 min, as per manufacturer instructions. Protein results were recorded in six categories which were: negative, trace, + (30 mg/dL), ++ (100 mg/dL), +++ (300 mg/dL), and ++++ (2000 mg/dL); pH results were recorded in seven categories: 5.0, 6.0, 6.5, 7.0, 7.5, 8.0, and 8.5; and leucocyte esterase results were recorded in five categories: negative, trace, + (small), ++ (moderate), and +++ (large). Cytologic evaluation of the uterine lavage samples was performed after cytocentrifugation (105 x g for 3 min) and staining using Camco stain Pak stain (Cambridge Diagnostic Products, Inc., Fort Lauderdale, FL, USA) by counting 200 cells (neutrophils, lymphocytes, macrophages, and uterine epithelial cells, excluding erythrocytes) and results were expressed as the percentage of total cells. Cows were considered positive for endometritis if neutrophils were >10% of total cells (Sheldon *et al.* 2006). The Holstein cows were between 40 and 60 days postpartum. All three (leucocyte esterase, protein, and pH) were increased in cows with cytologic endometritis and the associations were highly significant. Optimal cutoff points determined by receiver operating characteristic analysis for leucocyte esterase, protein, and pH were ≥ ++, 300 mg/dL, and pH ≥ 7.0, respectively. Combining the results for leucocyte esterase and pH improved the performance of the test strip, but this resulted in a group of cows (20.6% of the cows), which were approximately equally likely (46% with endometritis and 54% without endometritis) to have cytologic endometritis or not, and therefore could not be accurately classified. Reproductive impairment due to endometritis was restricted to multiparous cows; significantly decreased reproductive performance was observed for multiparous cows with lavage fluid leucocyte esterase ≥ +++ (154 *vs.* 115 median days not-pregnant), as well as cows with pH ≥ 7.0 (150.5 *vs.* 111.5 median days not-pregnant), but not in cows with high protein, even at the highest cutoff point. In conclusion, reagent strip test results were strongly associated with cytologic endometritis and reproductive impairment; however, in comparison with conventional cytology, the performance of reagent strip as an alternative test was relatively poor and needs further refinement.

The performance of leucocyte esterase in the latter study was not as accurate as described by Santos *et al.* (2006). While the sensitivity and specificity in the study of Santos *et al.* (2006) 48.8% and 73.3%, respectively, much higher corresponding values of 83% and 94% were reported by Cheong *et al.* (2012). The discrepancy between the two studies was attributed to variation in the cutoff percentage in PMN% and variation in the sampling period (1 and 7 wk postpartum by Santos *et al.* (2006); between 40 and 60 days by Cheong *et al.* (2012) Cows in the early postpartum period tend to have very high proportions of neutrophils in endometrial cytology (Gilbert *et al.* 2005), which may have improved the performance of the reagent strip test. The third reason for variation could be due to the time interval between sample collection and evaluation. The reactivity of leucocyte esterase in human urine with reagent strip is decreased in approximately 25% of samples after 24 h of refrigeration (Froom *et al.* 2000) increasing false negative results. In the study by Cheong *et al.* (2012), some samples were evaluated after the recommended period, inspite of which the false negative results remained low for leucocyte esterase. The sensitivity and specificity of pH to diagnose endometritis at the optimal cutoff point >7.0 was still relatively poor, at only 44.9% and 78.4%, respectively. The protein reagent strip test had the weakest association with cytologic endometritis and was not predictive of future reproductive performance.

x. Whiteside *Versus* Cytology *Versus* pH

Bedewy and Rahaway (2019) evaluated 42 repeat breeder cows and observed that subclinical endometritis had a significant association of PMN cells (12 to 18%) compared to 0 to 1% in the unaffected cows, with the increased pH value of cervical mucus (7.8±0.05) as compared with the unaffected cows (7.2±0.03). All the affected cows in this study were positive (100%) for Whiteside test, indicating subclinical endometritis. This obviously indicated the ability of Whiteside test in detection of subclinical endometritis (Raja *et al.* 2012). The increased pH in cervical mucus may be an outcome of presence of bacteria and inflammatory exudates (Salphale *et al.* 1993; Sheldon *et al.* 2006). An increased pH higher than 7.8 was a confirmation of endometritis (Singla *et al.* 2004; Bedewy and Rahaway, 2019).

y. Whiteside *Versus* pH *Versus* Bacterial Load

Kumar (2010) reported that the Whiteside test had a significant positive correlation with pH, bacterial load in cervicovaginal mucus, and bacterial load in uterine flushing.

Whiteside test and bacteriological examination were compared in normal and repeat breeder cows. The results of Whiteside test showed only 15% of control group had infection, but the remaining 85% were free of it. In contrast, 71.3% of the repeat breeders showed infection, while it was absent in 28.8% animals. In bacterial culture, 75.0% of the repeat breeding animals were found positive. The samples of control group that showed no color reaction in Whiteside test had also no growth in bacterial culture. Satheshkumar and Punniamurthy (2007) reported color reaction on Whiteside in 92.85% cervical mucus samples of normal animals indicating subclinical uterine infections. Krishnakumar *et al.* (2003) observed normal, mild, moderate and severe color changes in Whiteside of cervical mucus of 26 repeat breeding cows to be 0 (0.00%), 12 (46.15%), 10 (38.46%) and 4 (15.38%), respectively. Methai *et al.* (2005) recorded color changes in 100% samples of cervical mucus in repeat breeding cows suffering from endometritis. As the normal animals may also contain infections, some earlier workers could obtain complete sterile cervical mucus from normal breeding cows (Shukla and Sharma, 2005; Sharma *et al.* 2008). All the positive samples revealed single isolate (Sharma *et al.* 2008). Almost similar findings in normal breeding cows were reported by Das *et al.* (1996).

In conclusion, Sheldon *et al.* (2006) explained the difficulty in estimating sensitivity and specificity of the diagnostic methods of endometritis, given that it is hard to find a gold standard. One alternative method is reproductive performance, evaluated by diagnosing pregnancy at 120 d; however, this method will never achieve 100% sensitivity because many factors lead to infertility (Le Blanc *et al.* 2002a).

CHAPTER 3

THERAPEUTIC STRATEGIES FOR UTERINE INFECTIONS IN COWS

Pankaj Sood

Primarily, the uterine defense mechanisms occupy a pole position in dealing with recovery from uterine infections. In addition, the natural defense mechanisms also regulate the outcome of different extraneous agents used to treat uterine infection. However, to hasten the recovery, more so in severe forms, a variety of agents are used to treat uterine infections in cows. Broadly, these include uterine antiseptics, antibiotics and numerous other non-antibiotic alternatives. The following information discusses about different strategies in light of their solitary or combined use, the reproductive stage of cows treated, the success rates obtained in terms of different reproduction indices and limitations, if any.

A. NATURAL IMMUNITY: UTERINE DEFENSE MECHANISM

Effective defense against reproductive tract invasion by environmental organisms is mediated by anatomical and functional barriers as well as non-specific and specific immune responses (Dhaliwal *et al.* 2001).

The uterine defense mechanisms against contaminant micro-organisms are maintained in several ways (Asbury *et al.* 1980; Liu and Cheung, 1986):

- anatomically, by the simple or pseudostratified columnar epithelium covering the endometrium.
- chemically, by mucus secretions from the endometrial glands.
- immunologically, through the action of polymorphonuclear cells (PMNs) and humoral antibodies.

However, the degree of interaction among the aforesaid mechanisms is unclear. It is generally accepted that the cyclical pattern of steroid hormone concentration and characteristic for dif-ferent stages of the estrous cycle regulate the poten-tial

DOI: 10.4324/9781003749257-3

pathogenicity of microorganisms that contaminate the uterus postpartum. Nonetheless, disruption of the above mentioned mechanisms allow the opportunist pathogens, mostly the microbes present in the posterior gastro-intestinal tract and around the perineal area (Paisley *et al.* 1986; Hussain *et al.* 1990), to colonize the endometrium and cause uterine infection (Vandeplassche and Bouters, 1976).

a. Uterine Cellular Immunity

Inflammation of the endometrium can occur following coitus, AI, or more commonly in cattle after parturition. The PMNs, blood monocytes and tissue macrophages are considered as the 'professional phagocytes' in the cellular defenses against pathogenic micro-organisms. In the uterus, the cellular defense against bacterial contaminants is provided by uterine leukocytes (Vandeplassche and Bouters, 1976, 1983; Frank *et al.* 1983; Romaniukowa, 1984; Vandeplassche, 1984). After experimental intrauterine (i.u) infection, the PMN population within the uterine lumen usually increases (Watson *et al.* 1990; Butt *et al.* 1991).

During estrous cycle, the leukocyte numbers in the uterine lumen are relatively high during metestrus and diestrus compared to other phases. Importantly, their functional activity is unaffected by the stage of estrous cycle. In contrast, the leucocytic presence and activity differs in pregnant-postpartum compared to the cyclic cows. An inevitable microbial contamination of uterine lumen in a vast majority of the calved cows is evident 1 to 4 weeks postpartum, and some cows may even experience death even in the absence of direct trauma (Lewis, 1997). Guidry *et al.* (1976) found that numbers of peripheral blood neutrophils increased slowly from around 6 weeks prior to parturition to reach a peak on the day of calving but, conversely, maternal and fetal cortisol may suppress neutrophil function at the time of calving (Roth and Kaeberle, 1981, 1985; Whitmore *et al.* 1988). Immediately after calving, the phagocytic activity of blood neutrophils declines; within the uterine lumen, its cellular defense mechanisms are, however, maintained by increased PMN numbers (Guidry *et al.* 1976). The latter change is vital as around 48 h after normal unassisted calving, leukocytes accumulate in the uterine lumen, which marks the beginning of a normal cleansing and involution process of the uterus (Morrow, 1969). The cows with puerperal metritis also experience an initial decrease in phagocytic activity of uterine PMNs (Vandeplassche and Bouters, 1976). During the 1 to 3 weeks immediately after calving, the number of peripheral blood PMNs declines (Guidry *et al.* 1976; Kehrli *et al.* 1989; Gilbert *et al.* 1993), probably associated with their migration to the mammary gland and uterine lumen. Also, their phagocytic activity is reduced, this fall being more marked in older rather than younger cows (Kehrli *et al.* 1989; Gilbert *et al.* 1993).

This observation might explain why the incidence of postpartum disorders is higher in older cows rather than first and second calving heifers (Grohn *et al.* 1990). However, De Bois (1961) found that newly calved heifers were more susceptible to *A. pyogenes* infection than pluriparous cows, possibly associated with lower circulating immunoglobulin concentrations found in calving heifers. Subsequently, 2 to 3 weeks later, when clinical recovery has occurred, the phagocytic activity increases and this change coincides with lower numbers of bacteria in the uterine lumen (Vandeplassche, 1981).

The phagocytic activity of neutrophils in peripheral blood is greater than in the uterine lumen, in cattle (Chacin *et al.* 1990; Zerbe *et al.* 1996) as well as the mare (Vandeplassche and Bouters, 1983; Romaniukowa, 1984); on the contrary, Anderson *et al.* (1985) found no differences in the phagocytic activity of blood neutrophils compared to those found in uterine discharges. Across the domestic animal species, the differences in PMN phagocytic activity between blood and uterine discharges have been attributed to an unknown factor found in plasma (Vandeplassche and Bouters, 1983). Other leukocytes are present in the endometrium of all species. Interspersed within the endometrial epithelium in the basal region and sub-epithelial stroma, lymphocytes can be found in both cycling and non-cycling ewes (Lee *et al.* 1992) and heifers (Vander Weilen and King, 1984); their numbers do not vary significantly at different stages of the estrous cycle. In ewes, both non-granular and granular lymphocytes are found in caruncular and intercaruncular epithelium, and in lymphocytic nodules located within the stroma that also contain some plasma cells (Lee *et al.* 1988). High proportions of these lymphocytes are positive for the common leukocyte marker CD45, whereas CD5+ cells are scattered throughout the stroma. Within the basal epithelium and lymphocyte nodules, a small population of CD45R+ cells has been identified that are probably a small subset of T lymphocytes which do not express MHC class II antigens. These are 'natural killer' cells and have an important role as a first line of defense against invasive pathogens present in the uterine lumen (Lee *et al.* 1988). In cows, non-granular lymphocytes predominate (Lee *et al.* 1997), but their antigen expression has not been characterized. In the pregnant uterus, these granulated lymphocytes may control trophoblast invasion of the endometrium, particularly as their numbers increase in the intercaruncular endometrial epithelium during early pregnancy (Fox *et al.* 1998).

The cellular immune response through phagocytosis involves chemotaxis, adherence and attachment of PMNs to cell surface antigens on the microbe before it is ingested and finally digested by the phagocyte, which in the uterus may be adversely affected by several therapeutic strategies commonly used to treat postpartum disorders in cattle (Bouters and Vandeplassche, 1977). For example, manual removal of fetal membranes may inhibit uterine leukocyte phagocytic activity for several days (Vandeplassche and Bouters, 1983), as does i.u administration of most antiseptics and disinfectants (Vandeplassche, 1981, 1984). The i.u administration of antibiotics also suppresses uterine leukocytic activity; this could have serious implications in the treatment of metritis should the bacteria involved be resistant to a particular antibiotic. Conversely, several i.u chemotherapeutic agents may stimulate the uterine defense mechanisms; for example, Lugol's iodine and polyvinyl-pyrrolidone-iodine both cause necrosis of endometrial epithelium (Gustafsson, 1984) with subsequent release of prostaglandin F2α (Seguin *et al.* 1974; Gupta *et al.* 1983; Carleton and Threfall, 1984). However, some workers opine that the use of latter agents should be discouraged since they cause endometrial fibrosis (Nakao *et al.* 1988). To the contrary, we have found innumerable cows conceiving after infusion of 0.1 to 0.2% Lugol's iodine in cows with different grades of uterine infections (Sood, unpublished).

b. Uterine Humoral Immunity

Immunoglobulin concentrations in uterine secretions reflect both the extent of the endometrial inflammatory process in presence of microbial challenge and its chances of clinical recovery (Aknazarov, 1988). The immunoglobulins' of classes IgA, IgG and IgM are synthesized in the reproductive secretions of domestic animals and probably contribute towards local defense mechanisms and provide a protective role against pathogens (Curtain *et al.* 1971; Duncan *et al.* 1972; Whitmore and Archbald, 1977; Badinand, 1975; Dhaliwal *et al.* 1996a,b). Following experimental i.u inoculation of known pathogenic micro-organisms, immunoglobulins in cervical and vaginal secretions appear in the order IgM, IgA and IgG, and disappear in the order IgM, IgG and IgA (Corbeil *et al.* 1974). Their concentrations differ depending upon the site of sampling; IgG predominates in the uterine lumen and IgA in the vagina (Corbeil *et al.* 1974, 1976; Watson *et al.* 1990). Following systemic immunization, these immunoglobulins are undetectable in the reproductive tract (Butt *et al.* 1993).

Humoral antibody concentrations in the reproductive tract are stimulated following exposure to local antigen, and the response is site dependent. IgA is synthesized locally in the bovine uterus at the mucosal surface. IgG is derived from two sites: part of the IgG1 fraction is synthesized locally in the endometrium, while the remaining IgG1 and all the IgG2 are derived from the peripheral circulation (Curtain *et al.* 1971; Butt *et al.* 1993). Within the same species, the differences in immunoglobulin concentrations of uterine secretions have been attributed to sampling at different stages of the estrous cycle. The IgG concentrations are highest at estrus (Whitmore and Archbald, 1977), but few studies have investigated the specific influence of steroid hormones on immunoglobulin concentrations in genital secretions. IgG and IgM concentrations in lochia from healthy cows fall after calving (Georgiv, 1978). In cows with abnormal puerperium, both IgA and IgG concentrations in uterine fluids rise rapidly as endometritis develops; the concentration of IgA may end up being 2 to 3 times greater than in healthy cows, but IgM remains low (Aknazarov, 1988).

c. Influence of Steroid Hormones on Uterine Immune Response

Fluctuations in peripheral blood steroid hormone concentrations during the estrous cycle may influence uterine leukocytes, especially PMN numbers and their functional activity in uterine secretions. Histological examination of endometrium and the uterine lumen at different stages of the estrous cycle by different authors has proved a variation in the PMN count (Priedkains, 1987; Klucinski *et al.* 1990a,b; Ohtani *et al.* 1993; Faundez *et al.* 1994). In general, more PMNs are found in the uterus and its lumen during metestrus compared to other stages of the estrous cycle (Klucinski *et al.* 1990a,b; Faundez *et al.* 1994). Hawk *et al.* (1964) described a delay in leukocytes infiltrating the endometrium of heifers during the luteal phase of the estrous cycle following an i.u bacterial challenge. It has been hypothesized that during the estrogen phase of the ovarian cycle there is increased blood flow to the uterus, increased mucus production, and intensified PMN activity (Hawk *et al.* 1960, 1964; Killingbeck and Lamming, 1963); whereas in the luteal phase there is an absence of detoxifying agents in uterine secretions and a reduced endometrial epithelial permeability to bacteria that delays the leukocyte stimulation (Gunnink, 1973). It is generally accepted that the endometrium is more susceptible to infection under progesterone than estrogen dominance (Lewis, 2003; Vandeplassche, 1981, 1984). Interestingly, after i.u administration of oyster glycogen, a chemoattractant, resulted in a significant rise in uterine PMN numbers (Subandrio and Noakes, 1992, 1997) or a similar trend (Chacin *et al.* 1990) during the luteal rather than the follicular phase of the cycle.

Kerns *et al.* (1985) found an increased phagocytic activity of PMNs derived from the uterine lumen after treatment of ovariectomized cows with estrogen. Ovarioectomized cows, treated with progesterone, develop infection more frequently after experimental infection with *A. pyogenes* than those treated with estradiol (Carson *et al.* 1988). Similarly, Chacin *et al.* (1990) found an increased tendency for greater PMN activity in animals under estrogen influence, but there was no evidence that migration or activity of uterine PMNs was reduced significantly during diestrus or in progesterone treated ovariectomized cows. In line with the latter study, Anderson *et al.* (1985) found that the stage of the estrous cycle had no effect on phagocytic activity of uterine-derived PMNs, collected after i.u infusion of oyster glycogen.

A significant concentration of immunosuppressant proteins has been found in the uterine lumen that is capable of inhibiting lymphocyte proliferation during the luteal phase (Segerson *et al.* 1984). Chacin *et al.* (1990) found that treatment of ovariectomized heifers with progesterone enhanced the immunosuppressive activity of uterine secretions: a similar result has been achieved using a combination of estrogen and progesterone (Segerson *et al.* 1986). In addition, diestrus, or progesterone dominance, is unlikely to affect the migration rate of PMNs into the endometrium or adjacent lumen, or lower uterine resistance to contaminating micro-organisms; progesterone, instead may initiate synthesis of immunosuppressant proteins in the uterine lumen, which subsequently inhibit lymphocyte proliferation, rather than acting by itself as an immunosuppressant (Chacin *et al.* 1990).

Steroid hormones may also influence the functional activity of peripheral blood neutrophils, more so during estrus than diestrus (Killingbeck and Lamming, 1963). Whilst Roth *et al.* (1983) suggested that progesterone had a positive depressant action, Carson *et al.* (1988) observed that ovariectomised cows treated with estrogen, had higher phagocytic activity of circulating PMNs than untreated controls. To the contrary, Watson (1985) found no evidence that phagocytic activity of peripheral blood neutrophils was influenced by the estrous cycle. However, when neutrophils were opsonised with uterine secretions obtained from cows during the follicular phase rather than the luteal phase, there was a significant increase in their phagocytic activity. Using a sheep model, Ramadin *et al.* (1997) found that the inherent blastogenic ability of peripheral blood lymphocytes was increased during estrus. In addition, they found that (3H) thymidine was incorporated more efficiently into blood lymphocytes after ewes were infected experimentally *in utero* under estrogen dominance rather than in the presence of progesterone during diestrus. This supports the hypothesis that lymphocyte activity in cycling ewes is up-regulated by estrogens, but the suppressive effects of progesterone may override any estrogen influenced activity. This inter-relationship may play an important role in uterine immunological function.

Inhibition of lymphocytic proliferation can also be linked to progesterone as observed utilizing the studies in pregnant sheep and cattle. Liu and Hansen (1993) suggested that progesterone indirectly regulated T-cell function in the pregnant uterus by inducing secretion of endometrial proteins, described as uterine milk proteins (UTM-proteins), which are members of the serpin superfamily of serine protease inhibitors (Leslie and Hansen, 1991). The UTM-proteins have been localized by immunohistochemistry in endometrial glands, luminal epithelium and caruncular stroma of pregnant sheep (Moffatt *et al.* 1987). In sheep, ovine uterine serpin (OvUS) inhibits lymphocyte proliferation, perhaps mediating the direct inhibitory effects of progesterone on uterine immune function and reduced lymphocyte numbers in the endometrium (Gottshall and Hansen, 1992). The OvUS acts directly on lymphocyte cell surface receptors, but can only reduce lymphocytic proliferation through inhibition of peripheral blood killer cell-mediated events that include increased expression of CD25 in the absence of increased expression of interleukin-2: lymphocyte activation is down-regulated (Peltier *et al.* 2000). Whilst the uterine serpins have been identified in uterine secretions of both sheep and cattle after the first trimester of pregnancy (Moffatt *et al.* 1987; Leslie and Hansen, 1991), they have not been found in uterine luminal fluid of cycling and early pregnant animals.

The presence of uterine microbes vis-à-vis circulating steroids and neutrophil function in postpartum animals revealed some unexpected findings. For instance, *A. pyogenes* and *E. coli* in postpartum beef cows when progesterone concentrations were basal, did not cause uterine infection whereas, all cows developed uterine infections when the bacteria were infused after the onset of luteal function and progesterone concentrations had begun to rise. However, the mechanisms underlying these observations are not clear. For instance, Subandrio *et al.* (2000) showed that blood or uterine neutrophil function in cattle is not consistently up-regulated by exogenous estradiol, nor suppressed by progesterone. Plasma estradiol on the day of calving is also thought to contribute to the overall immune suppression observed in dairy cows around calving (Goff and Horst, 1997). Galvão *et al.* (2010a) also observed higher estradiol concentration on the day of calving for cows that developed uterine disease. The aforesaid discrepancy could have been answered by considering the concomitantly higher cortisol concentrations. High fetal and maternal blood cortisol concentrations associated with parturition may suppress circulating neutrophil activity and uterine defense mechanisms (Roth and Kaeberle, 1981, 1985; Whitmore *et al.* 1988; Kehrli *et al.* 1989). Cortisol is known to affect the pseudopodia formation in the PMN cells prior to chemotaxis. At calving, however, a reduced leukocytic activity through the influence of fetal or maternal cortisol, associated with the physiological control of parturition, may be compensated for by a rise in circulating neutrophils through the action of estrogen (Guidry *et al.* 1976).

From the above information, it can be concluded that besides PMN number or its activity, the cyclical pattern of steroid hormone concentrations, characteristic for different stages of the estrous cycle, also regulate the potential pathogenicity of micro-organisms contaminating the uterus.

d. Self-cure Relying on Inherent Uterine Defense Mechanisms

The prognosis in most cases of endometritis is fair to good. Many cases of endometritis are self-limiting and resolve after the resumption of estrus cyclicity and sexual rest. With each repeated estrous cycle the natural body defense mechanisms and the changes produced within the uterus and the genital tract by the estrogen tend to aid recovery. Thus endometritis is seldom static; nearly all cases recover spontaneously, as is the case with other inflammatory processes, with the passage of a variable period of time and subsequent estrus periods. However, the more severe the endometritis, the longer the time required for recovery. Slight endometritis may not prevent conception and from a biopsy the approximate length of time for recovery can be estimated (Brus, 1954; Studer, 1981).

Considerable self-cure rates (i.e., 15.6 to 55.0%) of cows diagnosed with acute puerperal metritis have been reported (McLaughlin *et al.* 2013; Sannmann *et al.* 2013b). In another study, the self cure rate was observed to be 23.52% (Sood *et al.* 2000). It is unclear whether low specificities of the diagnostic methods used to screen cows for acute puerperal metritis with a consequently high prevalence of type I errors (i.e., high false-positive rate) might contribute to this self-cure rate.

Because of the infectious nature, implications for animal well-being, and a potentially life-threatening course, treatment of acute puerperal metritis is considered beneficial rather than relying on the self-cure mechanisms (Beagley *et al.* 2010; Machado *et al.* 2012a).

B. ANTISETPTICS

An intervention for treatment of uterine infections is aimed to stimulate the uterine tone, blood flow and flow of the mucus. Earlier, especially in the postpartum cows, warm-to-hot, mild antiseptic solutions such as the aromatic soap solutions, mild quaternary ammonium, neutral iodophor, or chlorine solutions, physiological saline, dilute potassium permanganate solutions, sodium bicarbonate solutions, weak Lysol solutions, and dilute iodine or betadine solutions were used in 500 to 4000 mL amounts as vaginal douches. Mild, gentle massage of uterus transrectally 2 to 3 times a week may possibly be of value in producing similar effects. Applying 1 % acryflavine ointment on the external-os of cervix and cranial portion of vagina by massage, swabbing the cervix with Lugol's solution or other irritating iodine-base cervicitis mixture may also stimulate circulation, uterine tone, peristalsis, and the secretion of mucus by the endometrium. The i.u infusions or douching of uterus (with an objective of removing the contaminated material and debris) with an irritant antiseptic such as dilute iodine solution, 0.1% of Lugol's solution will strongly stimulate uterine muscle contraction and tone, greatly increase mucous secretions, will have an antiseptic effect and result in mobilization of leucocytes into the uterus. A similar sort of effect can also be produced by douching of uterus with acriflavine solutions, 1:500 to 1:3000 concentrations, chlorine, 200 ppm, chlorhexidine, or 200 ppm betadine solutions.

In cycling cows, weak Lugol's solution is quite irritating to endometrium. A 20 to 50 mL of 0.1% solution in distilled water or saline, into uterus, from 12 to 36 h post service has been found to be effective without reducing conception rates on the preceding or subsequent service, or causing luteolysis with a short 8 to 10 day cycle. The infusion of Lugol's or other mildly irritating solutions into uterus are liable to alter the length of estrous cycle. If the infusion in done during late diestrus, say at about day 15, the interestrus interval may be prolonged to 24 days. On the other hand, if the infusion is done on day 4 of the diestrus, the interestrus interval can be 8 to 10 days. The results of weak iodine solution have equaled or exceeded the results with other antibiotic or antiseptic products. Iodine is rapidly absorbed from the uterus. Its beneficial action may be due to microbicidal activity (antibacterial, antiviral, antimycotic), the local irritation, leucocytosis, and hyperemia; and it may correct iodine deficiency to some extent (Roberts, 1986).

a. Lugol's Iodine

Lugol's iodine (I_2KI) use is still widespread in Europe, although several studies have shown no benefits from this treatment on reproductive performance (Nakao *et al.* 1988; Tenhagen and Heuwieser, 1999; Heuwieser *et al.* 2000; Knutti *et al.* 2000).

The i.u infusions with various antiseptics such as Lugol's iodine and povidone iodine are relatively common for treatment for postpartum infections. The therapeutic efficacy of single infusion of 30 mL of 0.1 % Lugol's iodine at estrus in cycling cows with clinical endometritis was found to be 54.16% (Sood *et al.* 2000). In another study, Shukla and Pandit (1989) infused 30 mL of 0.25% Lugol's iodine for 3 days during estrus and reported a success rate of 65.45% in curbing subclinical endometritis in cycling cows. The discrepancy between two studies with different success rates can be attributed to low grade of uterine infection (subclinical) and repeated infusions in the latter study. On the other hand, relatively low success rates of 33.3% have been recorded by Baishya *et al.* (1999) after infusing 40 mL of 0.25% Lugol's iodine for 2 days starting from estrus in repeat breeding cycling cows with clear uterine discharge.

Apprehensions on the Lugol's concentration for i.u use exist among the practicing veterinarians. Although, 0.1% is the most common concentration used, a 2 days i.u infusion of 30 mL of 0.3% Lugol's during estrus and one day later, in cycling cows with moderate to severe clinical endometritis with intent of increasing the rate of clinical cures with an increase in Lugol's concentration was examined (Sood *et al.* 2012). The findings revealed a therapeutic success rate of 56.2%, which was similar to 54.16% obtained after single 0.1% Lugol's infusion as mentioned earlier (Sood *et al.* 2000). The cows not responding tended to have a greater magnitude of uterine microbial load before treatment (Sood *et al.* 2012). It was noteworthy that the 0.3% Lugol's iodine infusion for two days in the healthy cows revealed a normal endometrial histology in the succeeding estrus thereby suggesting the said protocol to be safe. However, 0.3% Lugol's resulted in discomfort to some cows after treatment. Therefore, with no additional clinical cure rates and from the perspective of animal welfare (discomfort), i.u infusion of 0.1% Lugol's is recommended.

Because i.u use of antiseptics may suppress the uterine defense mechanisms like phagocytosis, their use in the postpartum cow is not recommended. On the basis of the latter hypothesis, infusion of 30 mL of 0.1 % Lugol's iodine one day after estrus in cycling cows with clinical endometritis was investigated. The said protocol cured 71.42% cows with endometritis compared to 54.16% after infusion of similar volume and concentration of Lugol's during estrus (Sood *et al.* 2000). Higher success after Lugol's infusion one day after estrus was attributed to the fact that the uterine defense was allowed to work on the day of estrus and Lugol's infusion later performed its role the next day. Moreover, infusing the uterus during estrus may result in much of the infusion escaping from dilated cervix and contracting uterus (Seguin *et al.* 1974). Irrespective of the day of Lugol's infusion, the variables that could affect the treatment outcome were uterine tone, severity of infection and nature of infection (single or mixed). Mixed infection due to a competition for opsonization and hemolytic types of pathogens, are relatively difficult to be treated compared to the non-hemolytic ones (Sood *et al.* 2000).

The ready to use or stock solution of Lugol's iodine is stored in dark place or in amber colored glass bottles that are air tight. Investigating the sensitivity of Lugol's to light and air, it was observed that Lugol's iodine stored in transparent or dark glass containers for a period of 6 months and exposed to illumination intensity of 69.5 ± 2.3 lux (equivalent to 2 to 3, 40 Watts tube lights) did not affect the *in vitro* antibacterial activity. However, an air exposure for 72 h significantly reduced the optical density and *in vitro* antibacterial activity (Sood *et al.* 2012).

C. ANTIBIOTICS

Because of the infectious nature of uterine infections, antibiotics are considered beneficial for its treatment.

a. Pre-requisites

Culture and sensitivity test, preferably using a disc diffusion method is the foremost pre-requisite for antibiotic usage. Although the limitations of the latter include its qualitative nature (i.e., susceptible, intermediate, or resistant) rather than providing a minimum inhibitory concentration (Jorgensen and Ferraro, 2009) and the lack of mechanization or automation of the test. According to Rocksin *et al.* (2007), the broth micro dilution method is considered the method of choice for *in vitro* susceptibility testing of bacterial pathogens, which also provides data concerning the minimum inhibitory concentration (Pierce-Hendry and Dennis, 2010). Irrespective of the test applied, appropriate breakpoints are a prerequisite to differentiate susceptible from resistant bacteria and to guide a therapeutic approach based on clinical breakpoints (Supre *et al.* 2014). However, microbiological breakpoints derived from the minimum inhibitory concentration do not necessarily equate to clinical breakpoints (Dalhoff *et al.* 2009).

It is well known that the minimum inhibitory concentration of the pathogen is relevant for selection of an appropriate antibiotic drug (Sheldon *et al.* 2004a). In a meta-analysis, a minimum inhibitory concentration of the antibiotic drug applied was specified in 3 studies (13.0%). It is noteworthy that the antibiotic sensitivity of bacteria from uterine discharges has not been frequently tested (Malinowski *et al.* 2011). Hence, there is a significant need to encourage prudent use of antibiotics and alternative therapies to antibiotics (Ozawa *et al.* 2012). Such strategies involve a good knowledge of pharmacokinetics and contemporary information on the sensitivity of bacterial pathogens to be essential for clinical efficacy (Toutain *et al.* 2017) and selecting the most appropriate drug at its optimal dosage and duration to cure an infection while minimizing side effects and pressures for the selection of resistant strains (Fishman, 2006).

b. Factors Affecting Success of Antibiotics

i. **Route of administration (i.u *versus* systemic):** Route of antibiotic administration in cows with uterine infections has been a point of debate. The i.u route of administration of drugs for prevention and treatment of genital infections in cows has prevailed for many years. Initially, the agents used for such treatments were various disinfectants, such as Lugol's iodine, that had to be administered locally to act at the site of infection. However, with the availability of the antibiotics, systemic treatment was shown to yield considerable drug concentrations in several tissues and other body fluids. Hence, it became a common route for treatment of diseases in other organs than the genital tract, eg. the respiratory system. Likewise, it was demonstrated that certain cavities, like joints, could be reached by a systemic application of the antibiotic. Despite the latter advantage, the local i.u administration persisted to combat the uterine infections in the genital tract of large animals due to the reasons of an increased availability of the drugs meant for i.u use, an old belief of trust in its use and lack of information on the comparative efficacy of systemic *versus* local use of antibiotics. The i.u antibiotics get into the bloodstream, from where they can be taken up again by the uterus.

For antimicrobial treatment to be successful, an effective concentration of drug must be achieved and maintained at the site of infection for an adequate period. The uterine environment between a cyclic and postpartum cow is different. The environment of the postpartum uterus having endometritis is characterized by a low oxygen tension, presence of antibiotics-degrading enzymes, a mucopurulent discharge, and organic debris (Whitacre, 1992), which may diminish the efficacy of locally infused antibiotics. The uterine pathologic changes therefore results in poor absorption of the i.u administered drugs and adequate concentrations of the drug are not achieved in the sub-endometrial tissues, vagina and oviducts. Hence, there is a need that the selected antibiotic overcomes all the aforesaid limitations besides taking care of the microbial load. Systemic administration usually results in uterine tissue and lumen antibiotic concentrations equal to blood plasma levels (Gustafsson, 1980; Masera *et al.* 1980a; Bretzlaff *et al.* 1983a). For some drugs, the concentrations are same in the normal and pathologic uterus (Gustafsson, 1980; Bretzlaff *et al.* 1983b). Moreover, the systemic administration gives a better distribution of the antibiotics in the entire tubular genital tract and to the ovaries. Also, systemic administration eliminates the risk of damage to the endometrium and fetal membranes as well as abnormal exudates are unlikely to influence the drug distribution. Repeated treatment can be carried out in a relatively simple manner and without the introduction of new infections. Because there are reasons to assume that a moderate to severe uterine infection is seldom localized only to the superficial layer of the endometrium, therapeutic plans considering systemic treatment for distribution of the drug in the entire uterine investments might be beneficial. The available literature reveals that for the treatment of metritis, antibiotics are given by the i.u route, systemically, or both (Malinowski *et al.* 2011). Considering the aforesaid limitations of postpartum uterus, i.u treatment may be used as a complement, and a single i.u treatment would be justified only in very mild cases of endometritis. Irrespective of the route of administration, clearing of fresh postpartum uterus using contractile agents, such as oxytocin, may increase the effectiveness of the administered antibiotic.

ii. **Type of antibiotic:** Besides the limitations arising from an unfavorable uterine environment, certain antibiotics when used i.u are inappropriate for the treatment of uterine infections. For example, nitrofurazone is an irritant and has adverse effect on fertility. Aminoglycosides are not effective in the predominantly anaerobic environment of the infected uterus. Sulphonamides are ineffective because of the presence of para-aminobenzoic acid metabolites in the lumen of the infected uterus. Antibiotics such as penicillin, third-generation cephalosporins, and ampicillin have been systemically administered in conjunction with i.u instillations of oxytetracycline or ampicillin and cloxacillin for the treatment of cows with acute puerperal metritis (Nak *et al.* 2011). Comparing i.u and systemic administration, several antibiotics such as sulfonamides, tetracyclines, streptomycin, penicillin, ampicillin, gentamicin, chloramphenicol, are absorbed from the uterus. The absorption in the immediate postpartum period is considerably less than after complete uterine involution (Gustafsson, 1980). Presence of endometritis results in a further decrease of absorption (Gustafsson, 1980; Masera *et al.* 1980a, Bretzlaff *et al.* 1983b). The poor absorption results in high concentration of the drug in the uterine cavity, which may infact be advantageous in taking care of the infection of the superficial and inside layers of the uterus. However, the drug efficacy in the latter scenario may be affected by the anaerobic conditions *per se*. The reduced antibacterial effect may be due to a nonspecific interaction of drug with purulent material or a reduced uptake by sensitive bacteria any may therefore explain the poor efficacy of the i.u antibiotic therapy.

As shown for oxytetracycline and chloramphenicol, no significant differences were observed concerning the distribution of the drug between various tissues or between healthy and infected postpartum cows (Bretzlaff *et al.* 1983a, 1988). Hence, both the drugs may prove effective after systemic administration in the postpartum uterine infections. Broad spectrum antibiotics such as oxytetracycline used at a dose rate up to 22 mg/kg body weight provided effective minimal inhibitory concentrations in the lumen and uterine tissues. Systemic administration of penicillin resulted in genital tract tissue and lumen concentrations similar to blood plasma concentrations in cows. Other antimicrobials such as metronidazole, ciprofloxacin and cephalosporin are administered systemically as well as i.u for the treatment of uterine infections.

iii. Drug disposition depending on the route of administration

Systemic administration: In general, a drug after systemic administration is transported to the various body tissues where its concentration is influenced by several factors like plasma protein binding, tissue protein binding, lipid solubility, diffusion characters of the drug, dissociation rate constant and blood flow to the target tissue(s) (Sood, 1995; Sood, 2002a).

Penicillins: This has resulted in genital tract tissue and lumen concentrations similar to blood plasma levels in cow and mare (Arbeiter *et al.* 1976; Masera *et al.* 1980b; Gustafsson and Ott, 1981). A dose of 20000 to 25000 IU/kg body weight of sodium penicillin G (i.m) was found sufficient to combat most pathogens that are sensitive to the drug. However, the duration of desirable concentrations was relatively short (4 to 8h), which necessitates to repeat the treatment at least twice daily. The same dose of procaine penicillin gives a lower tissue concentration but longer duration and makes one daily injection sufficient. The systemic dose of semisynthetic or synthetic penicillins varies from 15 to 50 mg/kg daily to large animals (eg. ampicillin 15 mg/kg, i.m (i.m); ticarcillin and carbencillin 25 to 50 mg/kg, i.m). These doses of ampicillin and ticarcilllin give satisfactory concentrations in the uterine tissue (Lock, 1982). Bhatt *et al.* (1988a) studied disposition of ampicillin in normal and diseased (induced metritis) diestrus buffalo uteri following systemic i.m route (4mg and 6 mg/kg). The peak plasma levels appeared at 30 min but rapidly declined to non-detectable levels at 6 and 12 h post injection in normal and diseased animals, respectively. Irrespective of the dose and condition of the uterus, sufficient concentrations of drug were detected in the uterine tissues by 2h and with peak level at 4 h after administration. At 4 and 6 mg dose rates, the minimum therapeutic effective drug levels were maintained for 4 and 6 h, respectively, in both states of the uterus.

Jayachandran *et al.* (1990) revealed that after intravenous (i.v) administration of ampicillin (6 mg/kg), the peak concentrations of 17.81±1.25 and 1.09±0.10 µg/mL were attained at 15 and 120 min in plasma and uterine fluid, respectively. A therapeutic concentration of ≥0.1 µg/mL was maintained from 15 min to 8 h and 30 min to 6h in plasma and uterine fluid, respectively.

Aminoglycosides

Streptomycin: Jayachandran *et al.* (1988) elucidated the pharmacokinetics of streptomycin in healthy buffaloes after a single i.m injection of 10mg/kg. The drug was detected in plasma and uterine fluid for 30 and 12 h, respectively. The drug attained peak concentration of 24.39±2.67 and 5.06±0.18 µg/mL at 1 h in plasma and uterine fluid, respectively. The therapeutic concentration of the drug was maintained for 6 to 7 h in plasma and only for 1 h in the uterine fluid.

Gentamicin: Has a good *in vitro* effect against uterine pathogens with minimum inhibitory concentration values of 2 to 5 µg/mL (Ziv, 1980a). The i.m administration of 4mg/kg has been shown to give adequate concentrations for at least 6 h, while 2 mg/kg is too low a dose (Guedawy *et al.* 1983a). In a water vehicle, on the other hand, more than 50% of the infused gentamicin disappears from the uterine lumen within 1 h and almost all of it is gone by the end of the next five hours (Guedawy *et al.* 1983a). Therefore, gentamicin should be given in saline if disinfection is desired, like in a mild endometritis, while a water vehicle may be considered in more severe cases. It needs to be remembered that gentamicin is ineffective against anaerobes and has reduced activity in acidic or anaerobic environment. Haddad *et al.* (1985) administered gentamicin at the dose of 5mg/kg by i.m route to healthy mares every 8 h for 7 consecutive days. The biological half life and total body clearance were 2.13±0.43 h and 1.69±0.41 mL/min/kg, respectively. The mean endometrial concentration on day 4 and 7 were 5.02±3.3 and 12.75±1.6 µg/g. Haddad *et al.* (1987) studied the pharmacokinetic and tissue residues of gentamicin in cows at a dosage of 3.5 of 5 mg/kg. Mean half lives for absorption, distribution and elimination were 0.16±0.14, 2.59±0.53 and 44.91±9.38 h, respectively. Endometrial gentamicin concentrations ranged from 17.54 to 56.11 µg/g during the treatment with 5 and 3.5 mg/kg doses. Endometrial tissue gentamicin concentration exceeded serum concentration for all periods during the course of treatment.

Bhatt *et al.* (1988b) studied disposition of gentamicin (2 mg and 4 mg/kg) by i.m route in normal and diseased (induced metritis) diestrus buffaloes. The peak serum levels were attained after 30 min. The peak level of drug was significantly lower in metritis buffaloes as compared to the healthy ones. Gentamicin levels in uterus were above minimum inhibitory concentrations for 2 h in both groups of animals. On the contrary, Sinha *et al.* (1994) reported that gentamicin (10 mg/kg) by i.m route was unable to reach into endometrial tissue of healthy cows in sufficient quantity.

Amikacin: Brown *et al.* (1984) studied pharmacokinetics and endometrial concentration of amikacin (7 mg/g) after i.m route. The same dose was repeated at 12 h interval with a total number of five injections. The peak serum concentration of amikacin was attained after 1 to 2 h of each injection. The mean endometrial amikacin concentration was 2.5 µg/g. The values of elimination rate constant, volume of distribution and half life did not differ significantly after first and fifth injection.

Neomycin: Following i.m administration of sodium neomycin in chicks (20mg/kg), the peak serum levels were obtained after 30 min, which rapidly declined to low levels of 0.36μg/mL at 12 h. The minimum inhibitory concentration was maintained for 3 h only. While repeating the drug, it was found that the disposition results obtained after subsequent injection were quite comparable to those after first injection. Tissue distribution was homogenous except for very high levels in liver, kidney and lungs (Carli *et al.* 1982).

Baggot *et al.* (1985) administered neomycin (10 mg/kg, by i.v and i.m route) in horses and repeated the administration after 12 h by i.m injection. Pharmacokinetic analysis of neomycin revealed the peak serum levels of 23.2±10.2 μg/mL at 30 min, which declined to 2.8±0.8 μg/mL at 8 h. Neomycin was very rapidly absorbed after i.m administration site with an absorption half life of 0.16±0.05 h and systemic availability of 73.7±26.9 %. A peak tissue level, representing 40 % of total drug in body, was obtained 32 min post injection. At 8 h, the fraction of the dose in the central and peripheral compartment of the model was 1.5 and 2.5 %, respectively and 96 % of the cumulative amount eliminated upto that time.

Black *et al.* (1983) determined neomycin levels in bull calves following i.v and i.m administration at the dose of 12mg/kg. Following i.m administration the maximum serum levels was 14.5±6.0 μg/mL at 30 min. At 12 and 24 h after administration, the serum drug levels were 1.37±0.51 and 0.55±0.22 μg/mL, respectively. Burrows *et al.* (1987) compared the pharmacokinetics of neomycin in different age groups of calves following single i.v injection. In another study, following a single i.v dose of neomycin (5 mg/kg) in calves, the zero min intercept of the drug level was 45.03±1.49 μg/mL. The values for elimination half-life, volume of distribution (area) and clearance were 2.25±0.08 h, 0.24±0.01 L/kg and 1.27±0.04mL/kg/min, respectively. Therapeutic levels of neomycin persisted for 4 h after i.v administration. No detectable serum neomycin levels were observed at 10 h post administration (Gatne *et al.* 1993). In a subsequent study, the disposition kinetics, uterine tissue level and dosage regimen of neomycin were investigated in normal cycling buffaloes in estrus after its single i.v dose of 5 mg/kg. The distribution half-lives in peripheral compartment I and II were 0.07±0.00 h and 0.58±0.00 h, respectively, whereas the elimination half life was 10.81±1.59 h. The apparent volume of distribution and total body clearance were 0.56±0.00 L/kg and 37.0±2.8 mL/kg/h, respectively. At 10, 30 and 720 min the level of neomycin in uterine tissue were 42.43±2.84, 26.72±1.28 and 2.13±0.43 μg/g, respectively. Based on the kinetic parameters, a satisfactory i.v dosage regimen of neomycin in buffaloes would be 5 mg/kg followed by 3 mg/kg at 12 h intervals (Sood *et al.* 1999a).

The pharmacokinetics of the same dose and route of neomycin was found to be different in dystocia affected buffaloes. The kinetic parameters were determined by a three-compartment open model. The distribution half lives in peripheral compartment I and II were 0.09±0.01 and 0.73±0.13 h, respectively, whereas the elimination half life was 11.71±1.27 h. The apparent volume of distribution and total body clearance were 0.38±0.02L/kg and 23.6±2.2 mL/kg/h, respectively. At 10, 30 and 720 min the levels of neomycin in uterine tissue were 83.56±5.77, 49.07±2.39 and 3.56±1.54 µg/g, respectively. Based on the kinetic parameters, a satisfactory i.v dosage regimen of neomycin in dystocia buffaloes would be 3 mg/kg followed by 2 mg/kg at 12 h intervals (Sood *et al.* 1999b)

Chloramphenicol: Its use has been restricted to nonfood producing animals. In horses, therapeutic uterine tissue concentrations after i.u administration of 20 mg/kg lasted for less than 12h and treatment, therefore, had to be repeated twice a day to give desirable concentrations. The i.v administration of 20mg/kg results in acceptable tissue concentrations only for about 3 h. Bretzlaff (1986) studied the disposition of chloramphenicol in cattle. After continuous i.v infusion for 8h, a steady state plasma genital tissue ratio was approximately 3:1. In another study by Bretzlaff *et al.* (1988), chloramphenicol was administered to postpartum cows at a rate to reach a steady state plasma concentration of 5 µg/mL. After 6 h of constant i.v infusion mean plasma to tissue ratios were 3.05, 3.63 and 3.22 for caruncles, endometrium and uterine wall, respectively.

A study by Kochhar (1989) on chloramphenicol (30 mg/kg by i.m route) revealed peak concentrations of 12.75±0.018 µg/mL at 4 h as compared to 10.99±0.32 µg/mL in normal cycling buffaloes. The absorption and elimination half lives, area under curve, volume of distribution and clearance rates for normal and dystocia affected buffaloes were 1.14±0.01 and 1.19±0.13 h; 3.76±0.31 and 1.76±0.40 h; 86.41±2.71 and 121.73±10.04 mL/kg/h; 1.88±0.13 and 1.69±0.07 L/kg; 348.20±10.67 and 251.74±21.35 mL/kg/h. Uterine tissue concentrations were much higher than plasma levels at all the time periods.

Sulfonamides: This works well against *E. coli.* Sulfonamides are absorbed better by the postpartum uterus than other antibiotics (Gustafsson and Ott, 1981). However, sulfonamides do not work in the presence of pus and debris. Pharmacokinetic study of sulfadimidine (Singh *et al.* 1988) after its single i.v injection at the dose rate of 200 mg/kg revealed peak concentration of 314.0±13.0 and 100.2±2.5 µg/mL at 15 and 720 min in plasma and uterine fluid, respectively. The therapeutic levels were maintained for 24 h in plasma and 12 h in uterine fluid. Kochhar (1989) reported disposition of sulfadimidine

(100mg/kg) after i.m administration in dystocia affected buffaloes. The peak level reached 2h post injection. The values of area under curve, volume of distribution and body clearance were 2229.73±39.96 µg/mL/h; 0.16±0.03 L/kg and 41.85±4.99 mL/kg/h in healthy animals. The corresponding values in dystocia affected buffaloes were 2987.46±208.33/mL/h; 0.72±0.09 L/kg and 34.04±2.71 mL/kg/h. The levels of drug in the uterine tissue were manifold higher that the corresponding plasma concentration.

Oxytetracycline: Oxytetracycline remains among the antibacterial with alternatives in human and veterinary medicine for their indications and with a low risk for the spread of resistance (Anonymous, 2019). It is a pertinent antibiotic that has the ability to overcome the most microbial load of the uterus (Sheldon *et al.* 2004a), is a broad spectrum antibiotic that has the ability to work in an anaerobic environment and has an endometrium-restricted action, due to which it has been used for the treatment of postpartum endometritis for many years (Kaczmarowski *et al.* 2004). Earlier, oxytetracycline was reported as one of the antimicrobial agents approved by the FDA for use in lactating cows and thus specifically labeled for the treatment of endometritis caused by *Staphylococci* and *Streptococci* (Arrioja, 2001). Haimerl *et al.* (2017) reported that oxytetracycline and ceftiofur are among the most often used antibiotics in the treatment of metritis in cows.

The i.m administration of oxytetracycline resulted in a more rapid and complete absorption of drug into blood than after i.u infusion (Masera *et al.* 1980b). Bretzlaff *et al.* (1982) administered oxytetracycline (22 mg/kg by i.v route) in healthy cycling cows. After 8 h of attaining a steady state plasma concentration of 5 µg/mL the mean plasma and uterine tissue concentration of oxytetracycline was 4.86±0.68 µg/mL and 4.50±0.45 µg/g, respectively. The mean ratio of plasma to uterine tissue oxytetracycline concentrations was 1.08. In another study, following administration of oxytetracycline to healthy postpartum cows, after 8 h of attaining a steady state plasma concentration of 5 µg/mL, the mean plasma to tissue ratio were 0.95, 1.33, 1.88 and 1.04 for caruncles, endometrium, uterine wall and ovaries, respectively (Bretzlaff *et al.* 1983b). Bretzlaff *et al.* (1983a) failed to demonstrate the effect of endometritis on the distribution of oxytetracycline in the genital tract of postpartum cows.

In another study, the oxytetracycline pharmacokinetics after i.m administration in cows with clinical metritis associated with *Trueperella pyogenes* infection was investigated. The animals were treated i.m with long-acting oxytetracycline (20 mg/kg) and samples of plasma, milk and uterine secretions were collected for determination of the antibiotic concentrations. Pharmacokinetics of the antibiotic and *in silico* prediction of its penetration into the uterus was described. *Trueperella pyogenes*, with minimum inhibitory concentration values of 16 to 64 μg/mL was isolated from uterine secretions in some of the cows. Oxytetracycline showed fast absorption and penetration in the uterine secretions and milk. No change of withdrawal time for milk was necessitated in cows with clinical metritis. Maximum levels in uterine secretions and predicted concentrations of oxytetracycline in the uterus were lower than minimum inhibitory concentration values. Systemic administration of long-acting oxytetracycline did not guarantee clinical cure and was not a suitable choice for treatment of clinical metritis associated with *Trueperella pyogenes*. Hence, it was reiterated that the appropriate approach to antibiotic treatment of uterine infections of cows requires its adequate penetration at the site of infection and sensitivity of pathogens (Mileva *et al.* 2020).

Conventional *versus* nanoparticle oxytetracycline: The inability of oxytetracycline to eliminate bacterial endometritis completely in cows with clinical endometritis prompted a recent study to compare the efficiency of conventional and nanoparticle oxytetracycline in treatment of clinical endometritis in postpartum dairy cows (Ghallab *et al.* 2023). Oxytetracycline with different concentrations, oxytetracycline 5% (OTCC5%), oxytetracycline 20% (OTCC20%), and oxytetracycline 20% nanoparticles (OTC-NPs) (prepared from 20%OTC by high microfluidic homogenization for 6 h to result into particle's shape almost spherical, and their size, reduced 56.6 times to the original molecule, ranged from 1.271 to 1.924 μm with an average of 1.575 μm equal to 1575 nm, and the shape appeared rounded attached to each other by thin filaments) were used 3 times a week (once/week) by i.u route for the treatment of clinical endometritis. Diagnosis of clinical endometritis was based on the assessment of high serum concentration of pro-inflammatory cytokines, acute phase protein, increased endometrium thickness, and i.u discharges with different degrees of echogenicity monitored by ultrasonography. Application of OTC-NPs revealed a decrease in serum concentration of pro-inflammatory cytokines (IL-1, IL-6, and TNF-α) and acute phase proteins compared to OTCC20% and OTCC5% groups. The improvement achieved by OTC-NPs was attributed to the reduction of OTC particles into nano-size which facilitates its tissue bioavailability, dispersion, penetration power to deeper tissues, apart from having a broad-spectrum activity. These activities were clearly apparent after the evacuation of uterine contents using a single dose of PGF2α. The OTC-NPs revealed a reduction in serum concentration of cytokines compared to OTCC20% and OTCC5% groups arranged as follows: 10.11, 25.45, 35.56 for IL-1; 99, 300, 319 for IL-6; 1.01, 4.40, 8.06 for CRP; and 46, 183, 266 for TNF-α. Furthermore, an increase in serum concentration of albumin (3.34) was obtained by OTC-NPs compared to OTCC5% (1.70). This improvement can be taken as evidence of liver resumption functions and inflammatory reactions. On the other side, globulin concentration recorded an increase like albumin and total proteins in OTC-NPs compared to others. A reduction in the endometrium thickness in OTC-NPs with the disappearance of i.u discharges was monitored by ultrasonography. This confirmed the subsiding of clinical endometritis in OTC-NPs group. Moreover, a significant improvement in conception and pregnancy rate in OTC-NPs (77, 72%) compared to OTCC20% group (63, 63 %), and OTCC5% (50, 10 %), respectively were observed.

Intrauterine (i.u) administration

Penicillins: Following i.u administration, penicillin is rapidly absorbed with peak plasma concentration at 1 h and declining to less than 1 IU per mL at 4 h (Righter *et al.* 1975; Ayliffe and Noakes, 1978a; Masera *et al.* 1980a). In a comparative study, Masera *et al.* (1980a) reported that the peak plasma concentration after i.u infusion was low, reaching late and lasting for a shorter period as compared to i.m administration of the same dose. Endometrial concentrations, however, were considerably high upto 8 h with i.u route. It has also been found that administration of estradiol benzoate (10mg) in oil by i.v or i.m route before i.u administration of penicillin-G increased the absorption of sodium benzyl penicillin in endometritis suffering cows (Ayliffe and Noakes, 1982). The i.u infusion of natural penicillin (10 million IU sodium penicillin in a midcycle uterus) gives satisfactory concentration in the lumen and in the endometrium for upto 24h. Lower doses (1 million IU) are unlikely to yield desirable concentrations in a large postpartum uterus. If penicillins are used for i.u infusion to disinfect the uterine flora it appears that semisynthetic penicillins would be the choice since they are not absorbed alike the natural penicillins (Gustafsson 1980; Lock, 1982).

Bhatt *et al.* (1988a) reported that following i.u administration of ampicillin (2g and 3g); there was a poor systemic absorption which further decreased in the presence of metritis. The uterine tissue levels of the drug also declined in animals suffering from metritis. The duration for maintenance of therapeutic effective concentration of drug in uterine tissue was reduced to half of that in the healthy uteri. They recommended 3g ampicillin, i.u, repeated at 24 and 12 h interval, respectively for the treatment of unapparent uterine infection and metritis cases. A dose of 2g was incapable of sustaining effective tissue levels, for longer than 12 h and 6 h, respectively, in the two conditions.

Alam *et al.* (1990) studied uterine tissue and plasma disposition of ampicillin (1 g i.u) in healthy cows on the day of estrus. Peak uterine concentrations (16.17±0.66 µg/g) were recorded at 1 h with the therapeutic levels being maintained for 12 h. Following infusion, the drug appeared in plasma (0.26±0.01µg/mL) at 1 h with an average peak plasma concentration of 0.49±0.02 µg/mL at 2 h, which declined to 0.09±0.006 µg/mL at 4 h. The biological half-life of the ampicillin in uterine tissue was longer (4.5±0.16 h) as compared to the plasma elimination half-life in buffalo calves and lactating goats. The elimination half-life of ampicillin in buffalo calves and lactating goats have been reported to be 2.9±0.1 and 2.56±0.02 h, respectively (Alam, 1988).

Aminoglycosides

Dihydrostreptomycin: The i.u administration of 2.5 g dihydrostreptomycin at 2 to 3 days postpartum led to a peak concentration of 10 to 15µg/mL in serum of cows (Righter *et al.* 1975).

Gentamicin: Pedersoli *et al.* (1985) studied the endometrial and serum gentamicin concentrations in ovariectomised mares after its repeated i.u administration alone or in combination with estrogen or progesterone. The endometrial tissue concentrations at 24 and 120 h were higher when administered in combination rather than when infused alone. Serum gentamicin concentrations were lower than the endometrial concentrations. The peak level was recorded at 6 h post infusion in all the groups. In another study, following injection of 50 mL of gentamicin (5%, i.u) daily for 3 consecutive days in cows led to peak serum concentration (2.49±1.46, 6.60±5.47 and 4.98±2.70 µg/mL, respectively) at 3 h after each injection. The highest concentration of gentamicin in endometrial tissue (639.2±307.2 µg/g) was measured at 6 h post infusion decreasing to 9.6±3.6µg/g before next dose. At 71 h after last injection, the drug concentration in endometrial tissue was 0.86±0.43µg/g (Haddad *et al.* 1985). Bhatt *et al.* (1988b) reported that after administration of gentamicin (250 and 500 mg i.u) in metritic and normal buffaloes, the drug appeared in systemic circulation after half an hour. The peak uterine drug level was detected after 2 h which gradually declined to non detectable levels by 18 and 24 h, respectively. The minimum inhibitory concentration was maintained for 12 h in healthy uterine tissue, but only for 6 to 8 h in metritic tissue, irrespective of dose. Sinha *et al.* (1994) detected a maximum of 7.00±0.19 µg/g of gentamicin in endometrial tissue 6 h after its i.u administration at the dose rate of 5 mg/mg.

Neomycin: There are several reports that neomycin was used by i.u route to treat genital infections. Paufler (1969) reported an increase in the conception rate after instillation of Bykocillin-M (neomycin + penicillin) in diseased uterus. Fuquay *et al.* (1975) after inserting neomycin sulfate (1000 mg bolus, i.u) in normal postpartum cows found an increase in number of services per conception. Beggard *et al.* (1992) found neomycin-metronidazole combination more efficacious *in vitro* as compared to oxytetracycline for its activity against i.u flora of cows with retained placenta. After instillation of neomycin (3.3 mg/kg, i.u) in pony mares, very little neomycin was absorbed on day 2 compared to day 9 after foaling. Neomycin was not detected in any of the milk sample during the course of treatment (Boyd and Allen, 1988b).

Boyd and Allen (1988a) recorded absorption of neomycin from equine uterus following induced bacterial and chemical endometritis. Normal mares absorbed 6 and 56 % of neomycin in estrus and luteal phases, respectively, as compared to mares with bacterial endometritis which absorbed approximately 12 % of the drug in both estrus and luteal phases. Mares with chemical endometritis absorbed 31 and 64 % neomycin in estrus and luteal phases, respectively. The peak plasma concentrations occurred earlier in the infected mares compared to normal ones with maximum absorption in the luteal phase.

The plasma levels, pharmacokinetics and dosage regimen of neomycin were investigated in healthy and normal cycling buffaloes after its i.u administration (5 mg/kg). The drug was rapidly absorbed from the uterus. At 5 min 3.68±1.0 μg/mL of neomycin was detected in plasma. The absorption and elimination half-Jives were calculated to be 0.51±0.19 h and 5.87 ± 0.47 h, respectively. The apparent volume of distribution, total body clearance and bioavailability were 0.51 ± 0.02 L/kg, 86.7±4.4 mL/kg/h and 69%, respectively. Based on the pharmacokinetic parameters, a satisfactory i.u dosage regimen of neomycin in buffaloes would be 9 mg/kg followed by 7 mg/kg at 12 h intervals (Sood *et al.* 1998). In another study, the i.u dosage regimen of neomycin and its nephrotoxic effects were investigated in endometritis affected buffaloes. Neomycin was infused in the uterus of endometritis buffaloes at the dose rate of 5 mg/kg body weight during estrus. Very high bioavailability revealed that a large proportion of neomycin was absorbed from the uterus into systemic circulation. The plasma levels were utilized to calculate the pharmacokinetic parameters and subsequently the dosage regimen in endometritis buffaloes. An i.u dosage schedule of 7 mg/kg as initial dose followed by a single loading dose of 5 mg/kg body weight at 12 h interval was found to be most appropriate and free form any nephrotoxic effect (Sood *et al.* 2004). The *in vitro* erythrocytic penetration of neomycin in reproductively healthy and endometritis affected buffaloes is also on record. At blood concentrations ranging from 10 to 150 µg/mL, the mean neomycin penetration into erythrocytes of healthy buffaloes was to the extent of 20.62±2.19% of blood concentration, which was non-significantly lower than the value of 22.46±2.12% in the erythrocytes of endometritis buffaloes (Sood *et al.* 2002b). Neomycin penetration into erythrocytes of buffaloes of both groups was considered excellent as the value of packed cell volume in buffaloes is about 32 %. It has been reported that if a drug penetrates to the 3/4th extent of packed cell volume of a species the penetration is considered to be excellent (Srivastava and Malik, 1987). Antibiotics having good penetration in erythrocytes are preferred because erythrocytes store the drug, which is released slowly, and help in maintaining the therapeutic levels for a longer time (Ziv, 1980b).

Amikacin: Caudle *et al.* (1983) determined the levels of amikacin in the endometrium of mares in estrus after its i.u administration (2.0 and 3.0 g dose). Biopsy results after both doses indicated peak levels one h post infusion (mean values of 29.7 to 80.8 µg/g). The 3 g dose appeared to have no therapeutic advantage over the 2 g dose.

Chloramphenicol: Niedweski and Threlfall (1980) reported that chloramphenicol was rapidly absorbed from the postpartum (14 to 90 days) bovine uterus while the absorption from was lower (27.4%) from an early postpartum uterus. After infusion (5.5mg/kg), the peak serum concentration appeared at 30 min, which was greatly increased to 54.6% from day 60 postpartum onwards. Bretzlaff *et al.* (1988) studied distribution of chloramphenicol in the genital tract of healthy postpartum cows after i.u infusion (20 mg/kg). The mean concentration of chloramphenicol was 43.8 µg/g in uterine wall and 2.9µg/mL in plasma, at 8 h of injection.

Sulfonamides and trimethoprim: Bu (1955) reported higher levels of sulfonamide for a longer period in plasma during luteal phase than during follicular phase possibly because of reflux of infusate through cervix during follicular phase. Skhervan (1954) had demonstrated rapid expulsion of drug medium from uterus during follicular phase.

Bierschwal *et al.* (1955) infused 250 mL of 25% solution of sulfamethazine in the uterus of diestrus cows. The peak blood levels reached at 2 h and the therapeutic level was maintained for more than 12 h. Drug was present in measurable amounts even after 24 h of administration. In another study, a number of sulfonamides were shown to pass from the uterus into blood with peak concentration occurring at 2 h. Absorption was more pronounced in proestrus and estrus phase than in diestrus (Miller and Rouse, 1970).

Huang *et al.* (1972) detected endometrial concentration of sulfamethoxine for 24 h after i.u administration. Following i.u administration sodium salt of sulfamethoxine (66mg/kg; 25%) to cows on days 2, 6,10, 20 and 30 postpartum the therapeutic plasma level (≥50μg/mL) was maintained upto 24 h of administration in all groups of animals (Righter *et al.* 1975).

Chaudhary (1985) reported that both trimethoprim and sulfamethoxazole were detectable in higher concentrations for longer duration in plasma as well as uterine tissue during luteal than in follicular phase in the buffaloes with healthy or diseased genitalia. To the contrary, an increased absorption of sulfonamides (Miller and Rouse, 1970) and other antibiotics (Ayliffe and Noakes, 1978b) from the uterus has been demonstrated during follicular phase.

The disposition of sulfamethoxazole and trimethoprim combination (800 mg and 160 mg) was investigated in buffaloes after i.u administration. Both the drugs attained the peak plasma levels after 2 h of infusion. Uterine tissue levels at the corresponding time intervals and was maintained at 12 to 24 h. In endometritis induced buffaloes, the peak plasma concentration (μg/mL) of sulfamethoxazole and trimethoprim were 1.46 and 10.80 one h after infusion. Much higher concentrations of both the drugs (740.66 and 784.33 μg/g) were attained in uterine tissues as against plasma at 2 to 12 h post infusion (Chaudhary *et al.* 1987a,b).

In another study on pharmacokinetic profile of trimethoprim and sulfamethoxaxole (0.4 g plus 2.0 g, respectively, i.u) in normal and metritic buffaloes during luteal phase of cycle, it was found that both the drugs were quickly absorbed into systemic circulation with elimination half-lives of 1.5 and 2.6 h, respectively, for trimethoprim and sulfamethoxazole in normal and endometritis suffering buffaloes. The apparent volume of distribution was very low (Uppal *et al.* 1991).

Tetracyclines: Cohen *et al.* (1995) reported extra merits and benefits for clinical uses of oxytetracycline like endometrium irritation, activating uterine defense mechanism, inflammatory cells, and promoting i.u PMN infiltration, in addition to the renewal of endometrium upon its i.u use. Therefore, i.u infusion of oxytetracycline was a prevalent treatment of clinical endometritis, where the therapeutic concentration is maintained in the endometrium after treatment and the drug action is restricted to the uterine lumen and endometrium (Kaczmarowski *et al.* 2004; Roncada *et al.* 2000). In contrast, Cohen *et al.* (1995) reported that oxytetracycline couldn't penetrate uterine wall effectively and didn't eliminate uterine infection caused by *T. pyogenes*. Inspite of a major role in the anaerobic environment of the uterus after parturition, oxytetracycline has been used in postpartum cows (Kaczmarowski *et al.*

2004; Sheldon and Dobson, 2004). However, due to milk residues of the drug, use of oxytetracycline in treatment of endometritis seems skeptical. The use of i.u administered antibiotics has a restricted approval in the United States (LeBlanc, 2008). Several articles have described the pharmacokinetics of oxytetracycline after its i.u administration and studied its disposition in blood and milk (Roncada *et al.* 2000; Hajurka *et al.* 2003; Gorden *et al.* 2016). The i.u absorption of oxytetracycline is poor but residues can be still of concern (Lefebvre and Stock, 2012; Makki *et al.* 2016). After administration of 10% oxytetracycline solution once daily for three consecutive days, at a dose rate of 8 mg/kg body weight, by the i.u route, in cows showing clinical signs of grade I metritis (Sheldon *et al.*2009a) with purulent discharge from the uterus when examined on days 5, 10, 15 and 21 after parturition it was observed that the concentrations of oxytetracycline in the uterine secretion were 433.79 (39.17 to 1668.76) μg/mL six h after the third application at a dose of 8 mg/kg and 84.33 (1.58–467.55) μg/mL 96 h after the last treatment. These levels were higher than the minimum inhibitory concentrations-namely, between 4 and 64 μg/mL against pathogenic isolates *Trueperella pyogenes*, *Streptococcus intermedius*, *Escherichia coli* and *Bacillus pumilis*. Higher concentrations over time were measured in milk rather than in plasma, indicating the need to monitor the withdrawal time for milk.

Bierschwal and Uren (1958) observed rapid absorption of chlortetracycline into blood stream following i.u infusion (1 g crystalline chlortetracycline in 40 mL of aqueous solution) in diestrus cows. The peak serum level (0.05 to 0.18 μg/mL) was reported at 2 h of infusion. Righter *et al.* (1975) infused oxytetracycline in uterus of cows (500mg/kg) at 2, 10, 20 and 30 days postpartum. The time of detection of drug in plasma was inversely related to the days postpartum.

Bretzlaff *et al.* (1983a) infused oxytetracycline (5.5 mg/kg) in uterus of healthy and metritic postpartum cows. Mean values of the fraction of drug absorbed was 0.50 and 0.23% for healthy and diseased animals. At 24 h after infusion, oxytetracycline concentrations in caruncles and endometrium of all cows were higher than those in plasma, uterine wall and ovaries. The drug concentration in tissue of metritic cows was lower than those in the same tissue of healthy animals.

In a more recent study (Mileva *et al.* 2022) changes in oxytetracycline levels in the uterine secretion over time after i.u administration were determined in cows with metritis that were evaluated in relation to the sensitivity of the pathogenic bacterial isolates and their sensitivity to oxytetracycline. The concentrations of oxytetracycline in the uterine secretion were 433.79 (39.17

to 1668.76) μg/mL 6 h after the third application at a dose of 8 mg/kg and 84.33 (1.58 to 467.55) μg/mL 96 h after the last treatment. The i.u infusion of oxytetracycline for three consecutive days resulted in the successful treatment of metritis in terms alleviating inflammation and restoring the estrous cycle in cows.

Cephapirin: Cephapirin appears to be effective at improving fertility in animals with clinical endometritis (Hyde and Brennan, 2020). The effect of single i.u administration of cephapirin on clinical recovery, clearance of uterine bacteria and reproductive performance was evaluated in postpartum buffaloes (35 days postpartum) with subclinical endometritis (>10% PMNs in endometrial cytosmears). The treated animals were re-examined cytologically on the first postpartum estrus. The animals were considered to have recovered if their PMN cells dropped to ≤5% on first postpartum estrus. Presence of *Escherichia coli*, *Arcanobacterium pyogenes* and *Fusobacterium necrophorum* in uterus was detected based upon PCR amplification of genes related to bacteria specific virulence factors. A total of 66.7% of the treated buffaloes recovered as compared to 28.6% in the untreated control. Rate of bacterial clearance and median days to first service did not differ between treated and untreated animals. However, the treated buffaloes had significantly reduced days open (by 14 days). The treated buffaloes were as likely to conceive as normal buffaloes (Nehru *et al.* 2019). LeBlanc (2008) showed that cows receiving cephapirin on 27 to 33 days postpartum resulted in a 60% higher likelihood of becoming pregnant and 29% reduction in time to pregnancy compared to their untreated herd mates. Similarly, cows with subclinical endometritis that received cephapirin between 20 and 33 days postpartum had an 89% increase in the risk of pregnancy compared to non treated cows (Kasimanickam *et al.* 2005a) and cows treated with cephapirin at 41±14 days in milk were approximately 2 to 3 times more likely to be pregnant by 56 days into breeding season as compared to untreated control (McDougall, 2001). Kasimanickam *et al.* (2005b) reported that cows that were categorized as negative for subclinical endometritis using endometrial cytobrush cytology also showed some improvement in reproductive performance following treatment with cephapirin, suggesting that these animals may have had slight endometrial disease.

Rifaximin: Rifaximin is a semisynthetic antibiotic, belonging to the family of rifamycins. It was widely used at treatment of various diseases in human, such as acute diarrhoeal enteritis (Koo *et al.* 2012), small intestine bacterial overgrowth (Peralta *et al.* 2009; Hoffman *et al.* 2011; Gatta and Scarpignato, 2017), travelers' diarrhea (Flores *et al.* 2011; Alajbegovic *et al.* 2012;

Hu *et al.* 2012), irritable bowel syndrome (Kane and Ford, 2016; Li *et al.* 2016; Ponziani *et al.* 2016), ulcerative colitis (Guslandi *et al.* 2006), *Clostridium difficile* (Gawronska *et al.* 2017; Ng *et al.* 2019) and hepatic encephalopathy (Hudson and Schuchmann, 2019). Due to the characteristics of low-aqueous solubility, P-glycoprotein efflux, and low-permeability, rifaximin is considered as a non-absorbed oral antibiotic (<0.4%), a best choice for topical preparation. Recently, in veterinary medicine, rifaximin has been attracted widespread attention, which has been used in the treatment and prevention of animal diseases, such as bovine mastitis and endometritis, dermatological diseases, hepatic encephalopathy in pets. It is published that rifaximin has shown excellent clinical effects on the therapy of hepatic encephalopathy in dogs and cats (Lidbury *et al.* 2016). Polat *et al.* (2015) published the results on comparison of i.u ozone and rifaximin treatment in cows with subclinical endometritis.

Unlike several other antibiotics having issues of milk withdrawal, rifaximin has been considered to be safe (Table 4). Twelve healthy dairy cows were selected by random and treated by rifaximin i.u infusion at a dosage of 25 g/head (contain RIF 187.5 mg). Milk samples were collected before dosing (0 h) and at different time intervals (6, 9, 12, 18, 24, 36, 42, 48, 60, 72, and 96 h) after treatment. The results showed that the residues of rifaximin in the milk samples from 6 h to 96 h after administration were not detected, which were lower than limit of quantitation (1.0 ng/mL) and lower than maximum residue limits (60 ng/mL in milk). This study indicated that administration of rifaximin by uterine routine has a minimal distribution rate into the milk, and the rifaximin i.u infusion could be used in lactating cows with zero milk-withdrawal periods. Being a non-absorbed oral antibiotic (<0.4%), as mentioned earlier, retention of rifaximin into the site of administration may render it to be highly effect in treatment of uterine infections upon i.u administration (Liu *et al.* 2021).

Polat *et al.* (2015) investigated the effect of i.u ozone and rifaximin treatment in cows with subclinical endometritis in 53 Simmental cows with subclinical endometritis, diagnosed by ultrasonographic examination. The findings revealed that the interval (days) between treatment to pregnancy (46.4 +/- 6.2 *versus* 40.0 +/- 6.0), calving to pregnancy (129.4 +/- 9.0 *versus* 125.0 +/- 13.1), and insemination number (3.2 +/- 0.3 *versus* 3.1 +/- 0.5) after treatment were similar in both the groups. It was therefore concluded that i.u rifaximin treatment was as therapeutic as ozone.

Table 4: Comparison of rifaximin with other approved i.u infusion used for dairy cattle (Liu *et al.* 2021).

I.u Infusion	Specifications	Dosage	Withdrawal Time
Rifaximin	25g:187.5mg	Once	0 d
Cephapirin	19g:500mg	Once	0 d
Tetracycline hydrochloride	2g	Once every 24 h, 1~3 continuously Once every 72 h, 2~4 continuously	7 d
Oxytetracycline	25g:250mL	Once every 48 h, 3~4 continuously	3 d
Cefquinome sulfate	25g:0.9g	Once every 72 h, 1~2 continuously	7 d
Doxycycline Hyclate	24g:2g	Once every 72 h, 1~4 continuously	7 d

iv. **Dose, interval and uterine variables:** Irrespective of the route of administration, for antimicrobial treatment to be effective, it is essential that an effective level of drug be achieved and maintained at the site of infection for an adequate duration.

The dose of an antibiotic may not be constant and therefore needs to be adjusted to the microbiologic pattern and to other factors influencing the drug disposition. The manufacturer's recommendation is often based on a relatively high bacterial sensitivity. On the other hand, minimum inhibitory concentrations may vary on the type of microbe. For instance, 20µg/mL of oxytetracycline is required against *C. pyogenes* and 75µg/mL against *E. coli*. Furthermore, the *in vitro* validation of drugs against a microbe for their use in animals may be inappropriate *in vivo* where the natural defense mechanism works synergistically with antibacterial agent (Ziv, 1980a,b; Ball *et al.* 1984). Another variable that needs to be taken into account is the postpartum versus the cycling cows. The latter is justified when the recommended 10mg/kg systemic dose of oxytetracycline was to be increased to 11 mg/kg (twice daily) to maintain an adequate minimum inhibitory concentration of 5 µg/kg in the genital tissues (Bretzlaff *et al.* 1983b) and also had to be increased to 15 mg/kg (twice daily) in fresh postpartum cows. The increase in the dose of drug administration was to compensate the increase ratio of blood plasma to uterine tissue by 30% in postpartum cows. To the contrary, for i.u treatment with oxytetracycline, a total doses of 0.5 to 5 g has to been employed. A lower dose of 0.5 to 2 g is unlikely to yield adequate concentrations in the large postpartum cavity. The other problem of i.u administration in postpartum animals is the uneven distribution of the drug in the genital tract (Masera *et al.* 1980a; Bretzlaff *et al* 1983a) and failure of adequate concentrations in the sub endometrial tissues, in vagina and cervix or in ovaries and oviducts. In cows absorption of drugs seems to be greater under influence of estrogen (Gustafsson, 1980; Gustafsson and Ott, 1981). The possible adverse influence of the drugs on the uterine defense mechanisms of high concentration achieved locally in the uterus also needs to considered before deciding antibiotic treatment (Vandeplassche, 1981; Frank *et al.* 1983; Jayapa and Loken, 1983; Ziv *et al.* 1983). Another variable affecting the drug disposition is its molecular structure. For example, sodium benzyl penicillin is easily absorbed from mare's uterus (Gustafsson and Ott, 1981; Allen, 1978), whereas semisynthetic penicillins (eg. tricarcillin) are poorly absorbed (Lock, 1982).

v. **Duration of antibiotic treatment**:

Cephalosporins (Espadamala *et al.* 2018)

Ceftiofur hydrochloride: 2.2 mg/kg of BW for 5 days

Ceftiofur crystalline-free acid: 6.6 mg/kg of BW by s.c route, with a second dose after 72 h

Ceftiofur sodium: 1.1 to 2.2 mg/kg for 3 days

Ceftiofur has been approved for the treatment of acute puerperal metritis in the United States and Europe, specifically for a treatment for 5 consecutive days. Management practices associated with antibiotic use on dairy farms vary considerably and compliance is suboptimal (Sawant *et al.* 2005). Furthermore, anecdotal evidence from the field suggests that, on average, only 2 to 3 treatments are administered in treating feverish metritis. It is unknown if a shortened time above minimum inhibitory concentration causes suboptimal cure rates, poses a risk for subclinical endometritis, or reduces reproductive performance (Witte *et al.* 2011). Furthermore, an abbreviated treatment (<3 d) can be considered off-label drug use and, due to suboptimal pharmacokinetics and exacerbate emergence of resistant bacteria.

A meta-analysis by Haimerl *et al.* (2017) after specific exclusion criteria, 21 publications comprising 23 trials remained for final evaluation, which revealed that the majority of the studies (n=19) were attributable to the highest evidence level. Majority of the studies (n=17) applied ceftiofur for the treatment of acute puerperal metritis. Concerning the efficacy of ceftiofur, 7 studies observed clinical improvement, whereas none found improved reproductive performance. Fewer than half of the studies (n=10) performed a bacteriological examination and only 4 implemented an antibiotic susceptibility test. Also, 3 studies (13.0%) described a self-cure rate *per se*. Little attention was given to the issue of bacterial resistance (n=3), the need for reducing the application of antibiotics (n=2), or guidelines for prudent use of antibiotics (n=1). It was demonstrated that implementation of bacteriological examinations, sensitivity testing, and determination of minimum inhibitory concentrations, as well as reporting and discussion of critical issues (e.g., self-cure rates, resistance, prudent drug use), were suboptimal. Nevertheless, more high-quality research considering self-cure rates is necessary to address critical issues related to acute puerperal metritis and crucial to the dairy industry, such as resistance, prudent use of antibiotics, animal welfare, and cost-benefit ratios.

Another review on antibiotic treatment of metritis using antibiotics in dairy cows (Haimeri and Heuwieser, 2014) observed that the majority of the studies discussed the efficacy of the antibiotic applied based on clinical parameters such as cure rates as defined by fever, discharge, and, in some cases, haptoglobin concentrations or considering reproductive performance parameters. The evaluation on the basis of bacteriological cure rates, as frequently conducted in mastitis research (Kalmus *et al.* 2014), have been undertaken rarely. In general, the efficacy of antibiotic drugs is difficult to be evaluated conclusively from the literature as different routes, doses, and products were applied and different examinations were performed on farms by farm or study personnel or veterinarians. It is well known that not all infections are eliminated following antibacterial treatment (Malinowski *et al.* 2011), and definitions of self-cure and cure rates differ widely (Drillich *et al.* 2001; Galvao *et al.* 2009a). Ceftiofur (the most frequent drug investigated, with 17 trials), 12 and 6 studies assessed effects on clinical and reproductive parameters, respectively. Although 7 studies observed clinical improvement, none found improved reproductive performance after an application of ceftiofur compared with the reference group. Concerning the second most commonly applied antibiotic agent (oxytetracycline), only 1 of 5 studies evaluated the effect on clinical parameters. Evidence exists that higher doses than those indicated on the label are needed to maintain therapeutic tissue and blood concentrations of penicillin procaine (Prescott, 2000).

vi. Effect of vehicle: Gentamicin in saline has a much lower absorption from uterus in the cow than the same amount in a water vehicle. The i.u infusion of a total dose of 250 mg (0.5 mg/kg) in saline maintains adequate uterus lumen concentrations for at least 6 h. In a water vehicle, on the other hand, more than 50% of the infused gentamicin disappears from the uterine lumen within one hour and almost all of it is gone by the end of the next 5 h. Therefore, gentamicin should be given in saline if a disinfection of uterus is desired, like in mild endometritis, while a water vehicle may be considered in more severe uterine infections to achieve a deeper penetration of the uterine tissue with gentamicin (Guedawy *et al.* 1983a,b).

c. Effect of Antibiotics on Uterine Contractile Ability

Ocal *et al.* (2004) studied the effects of gentamicin sulfate on the contractility of myometrium isolated from non-pregnant cows. It was revealed that gentamicin sulfate inhibited spontaneous, as well as oxytocin or PGF2α induced contractions of myometrium in a dose-dependent manner. Although both the frequency and amplitude of contractions were significantly inhibited by gentamicin sulfate, the effect on the frequency of the spontaneous and agonist-induced contractions was more prominent than on the amplitude. The result from this *in vitro* study indicated that it may be important considering the potentially negative effect of gentamicin sulfate on uterine involution in cows with puerperal endometritis, resulting in impairment of fertility performance. The findings suggested a possible involvement of inhibition of membrane calcium channels in the inhibitory effect of gentamicin sulfate on myometrial contractions, as addition of calcium after establishment of maximum inhibitory effect of individual dose of gentamicin sulfate significantly restored the normal contractility (Paradelis *et al.* 1982). Similarly, macrolide antibiotics erythromycin (Celik *et al.* 2001, 2002) and clarythromycin (Celik and Ayar, 2002) have direct inhibitory effects on spontaneous and agonist-induced contractility of myometrium isolated from both pregnant women and rat (Granovsky-Grisaru *et al.* 1998).

d. Prophylactic Use of Antibiotics

The timing of antibiotic treatment in relation to uterine infection is vital; a variable timing of drug administration may have different outcomes. For instance, a study investigating oxytetracycline and flunixin treatment in cows with induced retained placenta and postpartum endometritis revealed that early oxytetracycline treatment of retained fetal membranes in the cows did not shorten the uterine involution or uterine infection and instead it slowed down the detachment process of the retained placenta. Oxytetracycline treatment after placental shedding might, however, shorten the uterine infection but otherwise did not affect the clinical results. Furthermore, flunixin treatment had no influence on the clinical outcome of the disease in the cows, which was attributed to normalization of fever and appetite at the time of flunixin treatment. An early treatment also changed the bacterial dynamics in the uterus, which was inferred as overgrowth of certain bacteria during and after treatment. Most importantly, following an early treatment with oxytetracycline, the placenta was still attached and infact postponed the expulsion of fetal membranes as this sort of approach continuously supplied the microorganisms with substrate for growth (Konigsson *et al.* 2001). Fecteau and Eiler (1996) showed that oxytetracycline does not interfere with the placental detachment process. Thus, an assumption would be that the process of placental shedding is favored by the bacterial infection as was proposed by Roberts (1986). To the contrary, although the cows that retained their placenta for a longer time, an early antibiotic-treatment in the cows with systemic illness had an early resumption of regained appetite and consequently had a higher mean energy intake per day than the other retained-placenta cows without any systemic illness (Konigsson *et al.* 2001).

e. Limitations of Antibiotic Therapy

Most practitioners perform antibiotic treatment of bovine uterine infections without sampling for bacteriological testing because it is time-consuming and costly. Such an empirical treatment strategy has been proposed as a risk for the development of bacterial resistance (Santos *et al.* 2010b). Santos *et al.* (2010a) found that over 50% of isolated *T. pyogenes* were resistant to amoxicillin, ampicillin, chloramphenicol, florfenicol, oxytetracycline and penicillin. In addition, 35% of isolated *E. coli* showed multidrug resistance (ampicillin–chloramphenicol–florfenicol) (Santos *et al.* 2010b). A subsequent study described a potential resistance to cephapirin, which is a commonly used antibiotic to treat cows with clinical endometritis in many countries (de Boer *et al.* 2015b).

Therefore each use of an antimicrobial drug, however, is associated with selective pressure for the emergence of resistant bacteria. Hence, there is a significant need to encourage prudent use of antibiotics and alternative therapies to antibiotics (Fishman, 2006; Ozawa *et al.* 2012). Recently, several publications have demonstrated emerging antimicrobial resistance of zoonotic organisms in food animals worldwide and expressed a potential threat to public health (Tragesser *et al.* 2006; CVMP,2009; Mann *et al.* 2011; Ozawa *et al.* 2012). It is obvious that increasing antibiotic resistance is associated with decreasing clinical efficacy and could exacerbate animal welfare and economic consequences.

- **Antibiotic residues in milk**: The global usage of antimicrobials in animals is double compared to humans (Aarestrup, 2012). Many studies have shown that significant portions (30 to 70%) of antibiotics are released unaltered, i.e., with potential antimicrobial activity, into the environment (Kümmerer, 2001). Upon release into the environment, most antibiotics are persistent and biologi-cally active (Sarmah *et al.* 2006). Milk is a highly consumed food item in the world which has also a great value for human health (Ventola, 2015). Hence, it is not surprising that a large proportion of administered antibiotics while treating infectious diseases in animals are present in their milk (Zhang *et al.* 2009). The developing countries are in greater risk due to residues in milk than the developed ones. Poor detection facilities as well as lack of proper monitoring system of residues in foods consider-ing the maximum residue limits might be taken as vital causes for higher risk of milk derived antibiotic resi-dues (Kebede *et al.* 2014). A meta-analysis of 364 studies in different regions of the globe revealed residues of β-lactam group to have been detected mostly 133 (36.54%), followed by tetracyclines, 51 (14.01%); fluoroquinolones, 49 (13.46%); sulfonamides, 46 (12.64%); and aminoglycosides, 38 (10.44%) (Sachi *et al.* 2019). It is also alarming to note that the antibiotics remain in an active form even after boiling of milk. In one such study, the thermal effect of certain antibiotic residues (oxytetracycline, amoxicillin, and ciprofloxacin) was evaluated in milk of cows that was boiled at 100^0C for 20 min. On day 2 after antibiotic treatment, 100% raw milk samples showed positive for antibiotics residue. After boiling, all milk samples were not only positive for different antibiotics residue, but as a matter of fact the antibiotic content was much more concentrated than that of the raw milk due to reduced water content on boiling. In the same study, no antibiotics residues were detected on day 9, which indicates the completion of the withdrawal period of the respective antibiotic (Anika *et al.* 2019). The potential effects of antibiotic residues in milk in terms of public health hazards are numerous (Beyene, 2016; Conzuelo *et al.* 2014) that include:

- **Antibiotic resistance:** Presence of low level of antibiotic residues in milk and other dairy products causes microorganisms to be resistant against anti-biotics. The resistant microbes may be transmitted among the individuals via direct contact or indi-rectly by exchange of resistant genes in the environment.
- **Allergic reactions:** Residues of various antibiotics are associated with multiple types of allergic reac-tions, including serum sickness and anaphylaxis, especially in case of penicillins.
- **Carcinogenicity:** Residues of antibiotics possess potential carcinogenic impacts by interacting with cellular elements, such as DNA and RNA.
- **Mutagenicity:** Mutagenic effect is another danger-ous impact of antibiotic residues, which can cause mutation of DNA molecule or damage of chromosomes. Infertility of human being may results from this mutation.
- **Teratogenicity:** Various congenital anomalies may be seen in new born child due to long term exposure of antibiotic residues during gestation period.
- **Disturbances in the normal intestinal environ-ment:** Normal habitant of the intestine coexists with others and colonizes to prevent the pathogenic microbes from producing diseases. Antibiotic residues in milk result-ing from

usage of broad-spectrum antibiotics may kill a wide range of microflora in the intestine including the non-pathogenic organisms, which can make the disease causing microorganisms more prominent and disrupt the normal intestinal environment.

In order to prevent inadvertent public health hazards due to antibiotics in milk, different strategies can be helpful. The first and foremost is adoption of milk withdrawal periods of the antibiotic treated cows. Withdrawal peri-ods of different antibiotics may vary from 1 or 2 days to couple of weeks. The minimum withdrawal period for oxytetracycline is 3 days (Haskell *et al.* 2003), for amoxicillin is 3 days (Martinez *et al.* 2014), and for ciprofloxacin is 6 days (Mahmood *et al.* 2016). In addition, previous studies have shown that withdrawal period may vary between different antibiotics, and the minimum withdrawal period maintained for the drugs was 7 days after the last dose of administration (Jayalakshmi *et al.* 2017). Drugs used for lactating and non-lactating ani-mals should not be intermixed, rather storing those in separate places. Marking of antibiotics treated cows for easy iden-tification, which will help the milkers to recog-nize them and withheld milk from marketing up to appropriate withdrawal time. Data regarding treatment of milking cows should be preserved cautiously in written form, where date and cause of treatment, name and dosage of drugs used, withdrawal time must be included. Antibiotics treated cows should be separated from the rest ones and milking lastly to minimize the risk of antibiotic residue contamination (Beyene, 2016). In addition, some antibiotics can be inactivated at different temperatures. Penicillin inactivates following refrigeration. Pasteurization can be used as an important measure to make most of the antibiotics inactive, while some others lose their activity if treated with UV radiation, activated charcoal or resin (Nisha, 2008).

Hauser *et al.* (1982) mentions the adverse effects of antibiotics on different components of the host immune response *viz.* chemotaxis, lymphocyte transformation, delayed hypersensitivity, antibody response. Though majority of the studies have been on the human subjects, the findings can be extrapolated in animal use as well.

Chemotaxis is the process by which phagocytes are attracted to the vicinity of pathogenic microorganisms by a number of factors including bacterial products, tissue proteases and complement components (Douglas, 1980). Chemotaxis is affected by gentamicin, tobramycin, amikacin, tetracycline, doxycycline, lymecycline, rifampin, and amphotericin-B.

Lymphocyte transformation is an *in vitro* technique in which lymphocytes are stimulated with antigens or mitogens to become metabolically active. Lymphocyte transformation thus measures the functional capacity of lymphocytes to proliferate following their exposure to antigens or mitogens (Yunis, 1973; Hoeprich and Martin, 1989). Delayed hypersensitivity may be thought of as an *in vivo* model of cell-mediated immune responses that is modulated by specifically sensitized thymus derived lymphocytes (T lymphocytes). The lymphocyte transformation is affected by trimethoprim, sulfamethoxazole, trimethoprim-sulfamethoxazole, amphotericin-B, tetracycline, doxycycline, minocycline, lymecycline, cephalothin, chloramphenicol, clindamycin and nitrofurantoin.

Delayed hypersensitivity is a cell-mediated immune reaction that can be elicited by subcutaneous injection of antigen with a subsequent cellular infiltrate and edema that are maximal between 24 and 48 h after antigen challenge (Wing and Remington, 1980). Amphoteracin B, metronidazole, doxycycline and tetracyclines have account for a delayed hypersensitivity in animals, whereas, rifampin has resulted in a similar effect in humans.

Antibody production has been affected by chloramphenicol, trimethoprim-sulfamethoxazole, rifampin and doxycycline.

Phagocytosis can be affected by tetracycline, doxycycline and amphotericin B. Sulfonamides, amikacin, gentamicin and tobramycin influence the microbicidal activity of phagocytes. Oxidative metabolism of PMN cell is known to be affected by chloramphenicol, trimethoprim, sulfamethoxazole, trimethoprim-sulfamethoxazole, and amphotericin B.

In addition, antibiotics such as oxytetracycline, as evaluated in the human RBCs, can cause a decline in their function of the antioxidant defense system, resulting in oxidative stress. Oxytetracycline also brought about morphological changes in human RBCs, leading to their hemolysis, when the oxytetracycline concentration was over 8×10"5 M (about 164 μg/mL) (Chi *et al.* 2014).

Another possible outcome of a prolonged intra-uterine antibiotic therapy and even leakage of antibiotics from uterus to vagina after intra-uterine therapy has been found to support the fungal growth as it disturbs commensal flora and reduce the competition for fungus by killing bacteria and changes pH of caudal reproductive tract (Zafracas, 1975; Chengappa *et al.* 1984). Nonetheless, a compromised genital defense mechanism is also a potential reason for fungal infestation of genital tract.

D. NON-ANTIBIOTIC ALTERNATIVES

a. Prostaglandins

Considering the limitations of antibiotics, PGF2α is one of the several non-antibiotic options recommended for treating endometritis in both cycling and non cycling cows, but its mode of action in non-cycling cows is not fully understood.

In cycling cows, myometrial activity is highest around estrus, which increases uterine contractility; it is lowest during the luteal phase (Al-Eknah and Noakes, 1989). This is associated with the inhibitory effect of progesterone on electrical conductivity in the myometrium. Uterine contractions are necessary for effective physical clearance of luminal debris through an open cervix and intracellular metabolites and inflammatory cells within the endometrium by lymphatic drainage (Guyton, 1991). In mares, several workers have showed that reduced uterine contractility contributes to the pathogenesis of persistent endometritis (Katila, 1996; Troedsson, 1999). Earlier, Troedsson *et al.* (1995) had experimentally infected mares using the i.u route and found that there was an increase in myometrial activity, probably associated with the release of prostaglandins from activated PMNs.

PGF2α is a naturally occurring luteolytic agent in non-pregnant ruminants. Oxytocin stimulates uterine secretion of PGF2α which, in turn, stimulates oxytocin release from the corpus luteum; this luteal oxytocin positively stimulates endometrial secretion of PGF2α. This positive feedback mechanism functions until luteolysis is complete (Cooke *et al.* 1998). As a result, peripheral blood concentrations of progesterone progressively fall from around day 16 of the estrous cycle until estrus, when they are undetectable. Conversely, estradiol concentrations rise to reach a peak when the animal shows standing heat. Hence, there are three reasons for using PGF2α for the treatment of endometritis. First, exogenous prostaglandin therapy administered to cattle presenting a functional corpus luteum will induce luteolysis and bring the animal into estrus, thereby removing the suppressive effect of progesterone on the uterine defense mechanism or, alternatively, stimulate it through estrogen. Furthermore, an increase in estrogen level and a reduction in progesterone result in myometrium contractions and an increase in PMN migration and phagocytosis (Azawi, 2008; LeBlanc, 2008). Hence, the luteolytic action of PGF2α has been used to make it the treatment of choice for endometritis in cows where a functional corpus luteum is present. This topic has been reviewed by Paisley *et al.* (1986) and Bretzlaff (1987). When the animal comes in estrum it can be additionally treated with antibiotics. Second, PGF2α stimulates myometrial contraction in cows (Lindell and Kindahl, 1983) and humans (Senior *et al.* 1992), which may be a mechanism that expels debris and micro-organisms that contaminate the uterine lumen after calving. Ulug *et al.* (2001) demonstrated that PGF2α stimulated the release of pro-matrix metalloproteinase-2 and pro-matrix metalloproteinase-9 from uterine tissue explants that are involved with breakdown of extracellular matrix mainly by degrading collagen type IV, which might aid myometrium contraction and uterine involution independent of cyclic status. Therefore after PGF2α treatment and the subsequently induced estrus in cows, cases of endometritis may resolve because of a mechanical clearance of debris and contaminating organisms from the uterine lumen, and by increased lymphatic drainage. Third, PGF2α may have a stimulatory effect on the phagocytic activity of uterine PMNs (Paisley *et al.* 1986). PGF2α stimulate phagocytosis of sheep erythrocytes by mouse macrophages (Razin *et al.* 1978). Whilst no similar studies have been carried out using bovine macrophages, such a stimulatory effect by PGF2α on uterine macrophages would enhance their role in the uterine defense mechanism, especially in non-cycling cows. Stimulation of phagocytosis by PGF2α can be attributed to increase in circulatory calcium levels after PGF2α administration in cows. In a study by Hernandez *et al.* (2014), the administration of PGF2α at the first hour and at 48 h after calving increased the serum calcium level and decreased the number of cases of subclinical hypocalcemia at day 7 postpartum. It also reduced the number of open days in dairy cows. Therefore, this treatment could be used to prevent reproductive problems in dairy cows. Increase in calcium after PGF2α administration might assist in the extracellular calcium uptake by leukocytes, which is a key initial feature of leucocyte activation and is necessary for cell proliferation and function (Lewis, 2001; Bréchard and Tschirhart, 2008). Leukocyte stimulation (i.e., interaction with antigen) also triggers a signaling cascade resulting in calcium influx from the endoplasmic reticulum to the cytosol. This increase in cytosolic calcium signals influx from the extracellular space via calcium release-activated calcium channels (Lewis, 2001). Given the large gradient between extra- and intracellular calcium stores (extracellular concentrations 800-fold greater than intracellular; Goff, 1999) and the relatively small number of leukocytes in circulation (Roland *et al.* 2014), it is unlikely that increased cellular calcium uptake contributes to decreased circulating concentrations (Waldron *et al.* 2003).

Immediately after calving, uterine contamination with micro-organisms interferes with its involution. The role of prostaglandins in this process is not well understood. After a normal calving, peripheral blood concentrations of PGF2α metabolites are high, and do not reach base line until about 14 days after calving (Guilbaut *et al.* 1984a); even with retained fetal membranes, this period may not be significantly extended (Kaker *et al.* 1984). The bovine uterus is the primary source of synthesis of the prostaglandin F series during the early postpartum period, particularly caruncular endometrium (Guilbaut *et al.* 1984a). The enzyme systems required to metabolize PGF2α are stimulated gradually as the rate of its synthesis increases (Guilbaut *et al.* 1984b). This related increase in synthesis and metabolism may occur also as a response to estradiol-17β treatment in cycling cows (Knickerboker *et al.* 1984). If the duration of PGF2α release postpartum is too short, involution is delayed; if this period is longer, the involution process is accelerated (Lindell *et al.* 1982; Madej *et al.* 1984; Fredriksson *et al.* 1985). The situation is different in cows with metritis or acute endometritis. Concentrations of PGF2α metabolite in peripheral blood remain elevated for longer than in healthy cows, and uterine involution is delayed (Kindahl *et al.* 1982; Lindell *et al.* 1982; Bosu *et al.* 1984; Watson, 1984; Fredriksson *et al.* 1985; Madej *et al.* 1986; Bolinder *et al.* 1988). High concentrations of PGF2α in the uterine lumen are indicative of inflammation (Kindahl *et al.* 1992). Lindell and Kindahl (1983) showed that administration of exogenous PGF2α postpartum promoted uterine involution in healthy cows. In general, anti-inflammatory drugs have no such effect (Odensvik *et al.* 1992). Guilbault *et al.* (1987) treated newly calved cows with 1 g flunixin meglumine twice daily for 6 days after their fetal membranes had been expelled, and found that peripheral plasma concentrations of 15-keto-13,14-dihydro-PGF2α were lowered. In these cows, uterine involution remained unaffected, but ovarian activity was reduced, compared to untreated cows or cows where PGF2α was administered concurrently with flunixin. This decrease in ovarian activity was characterized by significantly lower accumulated peripheral plasma concentrations of progesterone (Guilbault *et al.* 1984b, 1987).

It has been suggested that exogenous PGF2α is a suitable treatment for endometritis in the absence of a functional corpus luteum or in affected cows that have not been seen in-season. Any recommendations for treating endometritis in cows in the absence of a functional corpus luteum are based on clinical evidence alone (Young *et al.* 1984). There has been no evidence presented so far in cattle to show that uterine motility is significantly affected by prostaglandins, any increase has been for a short time only and was followed by a period of uterine refractoriness (Hopkins, 1983; Momont and Seguin, 1985). Exogenous PGF2α may enhance immune functions or increase uterine motility to help the uterus resolve infections in animals that do not have active corpora lutea (Nakao *et al.* 1997; Hirsbrunner *et al.* 2003). The results of clinical trials of PGF2α for treatment of clinical endometritis in the absence of an active corpus luteum are inconsistent (Steffan *et al.* 1984; Sheldon and Noakes, 1998; Le Blanc *et al.* 2002b). Steffan *et al.* (1984) showed that non-cycling cows with endometritis had fewer days open after PGF2α treatment than cows that received i.u antibiotic treatment or no treatment. This treatment may be successful because luteinized follicles that secrete small quantities of progesterone in postpartum cows prior to first ovulation are subjected to undergo lysis (Paisley *et al.* 1986), thereby removing the inhibitory effect of progesterone on the host's defense mechanisms. Le Blanc *et al.* (2002b) showed that injecting PG to cows at 20 to 26 days postpartum affected by clinical endometritis and with no CL on their ovaries was effective in management of uterine infections.

Several studies have described routine administration of PGF2α to healthy cows after calving. This initiated the early onset of ovarian activity, independent of the peripheral blood progesterone concentrations at the time of treatment and there was improved reproductive efficiency (Young *et al.* 1984; Etherington *et al.* 1988; McClary *et al.* 1989; Bonnett *et al.* 1990; White and Dobson, 1990; Glanvill and Dobson, 1991; Risco *et al.* 1994; Makki *et al.* 2017). Indeed, investigation by meta-analysis of many studies reporting the effect of PGF2α given postpartum indicated little benefit to the reproductive performance of dairy cattle (Burton and Lean, 1995).

The strategic use of PGF2α early postpartum, when purulent vaginal discharge and subclinical endometritis are prevalent (Gilbert *et al.* 2005), was initially thought to reduce the prevalence of subclinical endometritis, which could improve pregnancy rates per AI in cows inseminated exclusively by timed AI. Administration of exogenous PGF2α early postpartum promoted uterine involution in cows (Lindell and Kindahl, 1983). Accordingly, the effects of one or two treatments with PGF2α to reduce the prevalence of subclinical endometritis (presence of ≥ 5% PMN leucocytes in uterine smears) and improve pregnancy per artificial insemination in cows subjected to a timed AI program were evaluated. The two protocols utilized were either PGF2α treatment at 39 ± 3 days in milk (1 PGF2α) or 2 treatments with PGF2α at 25 ± 3 and 39 ± 3 days in milk (2 PGF2α). At 32 ± 3 days in milk, the prevalence of subclinical endometritis was reduced by treatment with PGF2a at 25 ± 3 days in milk in 2 PGF2α (control = 23.5% *versus* one injection of PGF2α = 28.3% *versus* two injections of PGF2α = 16.7%); however, this benefit disappeared at 46±3 days in milk, and 14% of the cows remained with subclinical endometritis. The disappearance of the beneficial effect of PGF2α was attributed to high spontaneous cure in the treated compared to control cows. One or 2 treatments with PGF2α did not influence pregnancy per AI on d 32 or 60 after timed AI, which averaged 39.9 and 35.2%. Similarly, treatment with PGF2α had no effect on pregnancy loss between 32 and 60 d of gestation (11.9%). It was concluded that 1 or 2 treatments with PGF2α before initiation of the timed AI program were unable to improve uterine health, pregnancy per AI, and maintenance of pregnancy in lactating dairy cows (Hirsbrunner *et al.* 2006). To the contrary, a single administration of PGF2α 14 to 28 days postpartum resulted in an enhanced first service conception rate (68% as compared to 43% in control cows) (Young *et al.* 1984). The advantages of this therapy, though, were restricted to herds with below average conception rates (< 50% to first insemination) (Young, 1989). Improved conception rate after single administration of dinoprost in cows 14 to 28 days post partum was particularly seen in cows with a serum progesterone value less than 0.5 ng/mL (Young *et al.* 1984). This fact demonstrated that the benefit of a PGF2α therapy was not the consequence of luteolysis (Young *et al.* 1984; Young and Anderson, 1986), but the beneficial effect of PGF2α administered post partum is rather the result of myometrial contraction and thereby accelerated uterine involution (Gustafsson, 1984; Garcia-Villar *et al.* 1987).

Interestingly, treatment of cows with PGF2α early postpartum may have a deleterious effect on the ensuing fertility. An unexpected negative effect on pregnancy rate of early (between 20 and 26 days in milk) treatment with PGF2α of cows that did not have a CL was recorded (Le Blanc *et al.* 2002b). It is biologically sensible that PGF2α would have no benefit in the absence of a responsive CL, and it is not clear how the drug produced an apparently harmful effect on fertility. However, Gay and Upham (1994) reported that administration of PGF2α at approximately 25 days in milk to clinically normal cows with a palpable corpus luteum significantly reduced first service pregnancy risk. They speculated that the treatment might have had an unspecified harmful effect on early follicular development. Similarly, Tenhagen and Heuwieser (1999) found that biweekly administration of PGF2α from 22 to 28 days in milk until first breeding at 50 to 60 days in milk reduced first service conception risk from 50% to 35%, in cows both with and without vaginoscopically diagnosed endometritis. The first postpartum ovulation typically occurs about 3 wk after calving in dairy cows and is followed by a short (<10 d) luteal phase (Stevenson, 1997). It has been reported that conditions including endometritis delay the process of re-initiation of cyclicity (Stevenson *et al.* 1988). It is possible that exogenous PGF2α during this critical period of transitional cycles may cause premature lysis of luteal tissue, depriving the hypothalamus, pituitary gland, or ovaries of progesterone needed to reestablish normal endocrine signaling and ovarian function. Practically, this reinforces the conclusion that cows do not benefit from treatment with PGF2α unless a CL is present. Given the imperfect ability to identify CL by palpation (Ott *et al.* 1986; Kelton *et al.* 1991), cows with or without endometritis should not be treated with PGF2α until at least 4 weeks postpartum.

In cycling cows with endometritis, ~49% recoveries to a single injection of PGF2α injected on day 7 of the estrous cycle were observed (Sood *et al.* 2000). The variety of microbes may vary at different locations. In one of our study, we observed 47 microbial pathogens comprising bacteria, fungi and an algae that were isolated and identified as single or mixed infection form 112 estrus cycling cows with clinical endometritis. Presence of *Brucella* spp. and *Prototheca* spp. in these cows was noteworthy (Sood *et al.* 2002c). In a subsequent study, the effect of microbial panorama on the therapeutic efficacy of single injection of PGF2α (injected on day 7 after estrus) in cows with clinical endometritis was evaluated. Out of 34.8% of cows that had infection from one particular microbe, 50% of these recovered, while the remaining 63.2% cows that had polymicrobial infection; 26.6% cows exhibited clinical recovery. Statistical difference in clinical recoveries between cows having single and mixed infection was significant. Recketts and Mackintosh (1987) postulated that in mixed infection the efficiency of natural defense mechanism is mitigated as a result of competition for opsonins that are essential for phagocytosis. This might be the plausible reason for fewer clinical recoveries in cows with a polymicrobial infection. Another aspect of this study was resistance of a particular microbe to PGF2α therapy. Barring *Corynebacterium* spp., *E. coli* and *Candida* spp. other microbes did not offer much resistance to PGF2α (Sood *et al.* 2003). Smooth strains of *E. coli* as recorded in extraintestinal infection have a hydrophobic surface, which tends to resist phagocytosis (Charles, 1988). As regard to *Candida* spp., the first line of defense against *Candida* includes activation of the alternative pathway of the complement system for attraction of neutrophils and subsequent engulfment of invading hyphae or pseudohyphae by neutrophils. Nonetheless, due to the size of *Candida* spp. neutrophils cannot totally ingest them (Tizard, 1996). It is again recalled that stimulation of phagocytosis by uterine leucocytes is the major pathway through which PGF2α eliminates uterine infection

Combined treatment with PGE2 AND PGF2α

The effect of PGF2α and PGE2, PGF2α alone or placebo was evaluated. The results of this study indicated that the administration of PGF2α or a combination of PGF2α and PGE2 21 to 35 days post partum had no beneficial effect upon measured fertility variables. The exception was a tendency for a shorter interval from calving to first insemination after administration of the combination of PGF2α and PGE2, as compared to the placebo group. The logic behind a combined administration of PGF2α and PGE2 was to harness a cumulative effect related to uterine pressure, if PGF2α and PGE2 were given in combination (Hirsbrunner *et al.* 2003). In humans, the potency of PGE2 in stimulating smooth muscle contractility exceeded that of PGF2α (Giannopoulos *et al.* 1985), whereas the two substances were equally potent as to myometrial contractility in cows near parturition (Zerobin *et al.* 1973). In experimental cows, increased uterine pressure of PGE2 compared to PGF2α was only measurable within the first 15 min after administration (Hirsbrunner *et al.* 2003). Obviously, there are two different receptors for PGE2 and PGF2α, but they also seem to be able to bind to the same receptor with different affinities (Bauknecht *et al.* 1981; Giannopoulos *et al.* 1985). PGE2 may even bind to the PGF2α- receptor with higher affinity than PGF2α itself (Giannopoulos *et al.* 1985) and furthermore, PGE2 is degraded via PGF2α (Canete Soler *et al.* 1987). In humans, it was also suggested that a greater number of PGE receptors were responsible for the greater potency for PGE's in stimulating uterine contractions compared to that of PGF2α (Hofmann *et al.* 1983). There are copies of the natural PGF2α and synthetic more potent cloprostenols which are racemic or in the d-form. The d-cloprostenol (as compared to PGF2α) can be used in a lower dose than the l-cloprostenol, and it has an equal effect on myometrial contractility (Hirsbrunner *et al.* 1998).

b. Immunomodulators

Much of the original research associated with immunomodulators of the uterine defense mechanism and their application as treatments for metritis or endometritis has been carried out in mares (Blue *et al.* 1982; Mitchell *et al.* 1982; Asbury, 1984). Thereafter, Anderson *et al.* (1985), Hussain and Daniel (1992) and Subandrio and Noakes (1992, 1997) reported similar work in cattle.

Immunomodulators were considered as an alternative approach to antibiotics by virtue of their inconsis-tent results for treating uterine infections. Immunomodulators activate uterine defense mechanism and when infused into uterus initiate local immune system (Dhaliwal *et al.* 2001). Some of the immunomod-ulators used in the treatment of endometritis are as follows:

i. E. coli lipopolysaccharide
ii. Oyster glycogen
iii. Leukotriene B4
iv. Granulocyte-macrophage colony-stimulating factor
v. Interleukins
vi. Serum, plasma, or hyperimmune serum
vii. PMN extracts and their components
viii. Bacteria-free filtrate

i. **E. coli lipopolysaccharides:** Serotype 026:B6 of *E. coli* lipopolysaccharide has been used after dissolving its 100 μg of it in 20 mL phosphate buffered saline to be administered i.u on the day of estrum. When placed in the uterine lumen by infusion, *E. coli* lipopolysaccharides are thought to act as a chemoattractant to PMNs. This increase in numbers of PMNs in the endometrium may help to resolve endometritis in both cows and mares (Asbury and Hansen, 1987; Williamson *et al.* 1987). A preliminary study by Targowski (1984) showed that i.u infusion of 100 µg *E. coli* lipopolysaccharide in healthy cows increased the PMN number in uterine secretions 6 and 24 h after administration. Similar results were obtained after lipopolysaccharides treatment of both healthy cows and those where endometritis was experimentally induced (Hussain and Daniel, 1992), which suggested that this may offer an alternative therapy to the traditional use of i.u antibiotics, or parenteral prostaglandins. A single i.u infusion of 100 µg *E. coli* lipopolysaccharides at estrus in repeat breeding cows cleared the bacterial infection from the uterine lumen within one estrous cycle. A 100 fold increase in total leucocyte count with more than 80% neutrophils having phagocytic activity of five bacteria per neutrophil in uterine flushing collected after lipopolysaccharides were reported. Further, a single intra-uterine infusion of 100 ìg *E. coli* lipopolysaccharides was effective in clearing the bacterial infection (75% of the treated cows with no bacterial growth in the estrus succeeding treatment) from uterine lumen within one estrous cycle. A 100% pregnancy rate in the 12 treated cows was recorded (Singh *et al.* 2000). Following this treatment, a majority of the repeat breeding cows with turbid vaginal discharge conceived as reported by another study (Dhaliwal *et al.* 1998). Deori (2002) observed that influx of PMN cells increased by about 77%, 63% and 68% and along with conception rate of 83%, 67% and 50% following intra-uterine infusion of standard lipopolysaccharides, indigenous lipopolysaccharides and bacteria free filtrate, respectively in endometritis cows.

Administration of i.u 100 µg E. coli lipopolysaccharides in both healthy and experimentally induced endometritis cows neither altered peripheral white cell count nor caused any systemic illness (Hussain and Daniel, 1992). Similar results have been described following i.u administration of lipopolysaccharides in ewes (Elmore *et al.* 1983) and mares (Couto and Hughes, 1985; Mollett *et al.* 1985). When *E. coli* endotoxin was administered to cows 5 days after calving at rates of 5µg/kg body-weight, the toxin was absorbed with associated transient clinical signs; however, when given 20 days later, there was no such effect (Peter *et al.* 1990). The contaminating micro-organisms associated with uterine infection postpartum may themselves secrete endotoxins detectable in peripheral blood, and these may interfere with ovarian function characterized by short cycles in cows (Peter *et al.* 1987). *E. coli* endotoxins may be absorbed from the uterus in newly farrowed sows (Elmore *et al.* 1978). Variation between species in endotoxin absorption from the uterus postpartum may be associated with differences in placentation and the extent of trophoblast invasion of the endometrium. In species, where there is little clinical or physiological evidence of endotoxin absorption, an intact endometrium may act as a cytological barrier to totally or partially exclude the toxin (Hussain and Daniel, 1992). This may explain the different outcomes of i.u endotoxin absorption in cows seen by Peter *et al.* (1990) immediately postpartum, where the endometrial epithelium is repairing after placental separation, there was a clinical effect. On the other hand there was no effect in the cycling cows.

ii. **Oyster glycogen:** PMN migration into the uterine lumen of healthy cows is stimulated after i.u administration of oyster glycogen, up to 90% of all cells identified in uterine secretions being neutrophils (Anderson *et al.* 1985; Chacin *et al.* 1990; Subandrio and Noakes, 1992, 1997). Variable concentrations of oyster glycogen ranging between 0.1 and 10%, all in 60 mL of vehicle, produced identical responses with a peak in PMN concentration 12 h after administration (Anderson *et al.* 1985). Measurable IgG concentrations were found in uterine secretions following treatment, but no IgA. Similar results were obtained in ovariectomised cows following both exogenous progesterone and estrogen administration (Chacin *et al.* 1990).

   A single i.u infusion of 500 mg oyster glycogen increased the uterine neutrophilic influx, cleared bacterial infection and improved conception rate by 66.67% (Singh *et al.* 2003). A similar study done by Prasad *et al.* (2009) in cross bred cows with bacterial endometritis resulted in an overall conception rate of 75%. Krishnan (2011) found 75% recovery rate along with increase in conception rate (41.67%) using 500mg single i.u infusion of oyster glycogen.

   In cows with subclinical endometritis with a history of repeat breeding, a single i.u infusion of 50 mL of 10mg/mL of oyster glycogen in phosphate buffer saline resulted in a recovery rate of 75.0%, with 50.0% of the recovered cows getting pregnant compared to 10.0% in the untreated control. Furthermore, the oyster glycogen treatment resulted in a significant decline in inflammatory mediators *viz.* lipid peroxide and nitric oxide from a pretreatment level of 753.73±32.78 nmol/L and 76.67±4.59 ìmol/L to 630.88±31.16 nmol/L and 60.08±2.17 ìmol/L, respectively (Krishnan *et al.* 2015).

iii. **Leukotriene B4:** Leukotriene B4 is another effective chemo-attractant, stimulating preferential migration of PMNs into the lumen of the bovine uterus. It is an acrachadonic acid metabolite known to be physiologically produced by stimulated granulocytes, macrophages and mast cells and plays an important role in neutrophil accumulation during inflammation (Zerbe *et al.* 1996). It stimulates the phagocytic capacity of PMNs particularly, in bovine (Hoedemaker *et al.* 1992) and may also stimulate the activation, proliferation and release of cytokines by B and T lymphocytes, monocytes and natural killer cells (Claesson *et al.* 1992). Injections of PGF2α may also work in part by increasing the uterine production of Leukotrine B4 and seem to promote uterine immune defense, as well as uterine health (Lewis, 2004).

A single i.u treatment of a 30 nmol/l solution increased the i.u leukocyte count 5 to 10 times within 24 h. Before stimulation, PMNs comprised 25 to 30% of the total leukocyte count, which after treatment increased to 85% (Zerbe *et al.* 1996; Dhaliwal *et al.* 2001). In another study, Krishnan (2011) assessed the recovery rate by using 50 mL of leukotriene B4 at the concentration of 30 nmol/L on the basis of negative to color reaction to the Whiteside test, normal pH of cervicovaginal mucosa and the presence of<5 % PMN cells in uterine cytology smears and observed a significant increase in recovery rate (83.33%) of endometritis cows. In cows with subclinical endometritis with a history of repeat breeding, a single i.u infusion of 50 mL of 30 nmol/L leukotriene B4 exhibited a recovery rate of 83.3% and increase in conception rate (50.0% in treated versus 10.0% in the control group). Compared to control, the cows in treated groups had greater leucocytic count at 24 hr following treatment. The infusion of leukotriene B4 compared to control was associated with a significant decline in the plasma levels of inflammatory mediators (lipid peroxide and nitric oxide) from the day of treatment (667.15±42.85nmol/L and 71.58±2.57 μmol/L) to subsequent estrus (555.56±42.53 and 60.08±2.17 ìmol/L), respectively (Krishnan *et al.* 2015).

It needs to be remembered that Leukotrine B4 is sensitive to light, temperature and oxygen (Zerbe *et al.* 1996) and appropriate measures need to be undertaken to maintain its stability.

iv. **Granulocyte macrophage colony stimulating factor:** Granulocyte macrophage colony stimulating factor, which had been genetically engineered, increased the PMNs in circulations by 12-fold and there was good scope for exploring the use of this factor for the treatment of uterine infections in cows (Hughes and Couto, 1988). Prophylactic use of bovine granulocyte-colony stimulating factor has the potential to preemptively increase neutrophil numbers and modulate their function ahead of parturition (Kehrli *et al.* 1991). Pegbovigrastim (Imrestor Elanco Animal Health, Greenfield, IN), a form of granulocyte-colony stimulating factor covalently bound to polyethylene-glycol to increase its half-life, is effective in maintaining increased neutrophil granulocyte levels, compensating for decreased bacterial killing (Kimura *et al.*

2014). In addition, pegbovigrastim has been shown to have an influence on gene expression in neutrophils, affecting gene families related to neutrophil function, migration, interaction with pathogens, and cellular survival (Heiser *et al.* 2018). Administration of pegbovigrastim 7 days before and within 24 h of calving, effectively improved granulocyte production and function *in vivo*, as well as in milk. The results showed a marked increase in the total concentration of granulocytes and monocytes as early as 2 d after the first injection, and this increased between the samples taken 2 d after calving. No migratory or killing differences were identified between granulocytes of treated and untreated groups, suggesting that pegbovigrastim-induced granulocytes were as effective as non-induced cells (Tombácz *et al.* 2019).

v. **Interleukins:** Zerbe *et al.* (2003) have studied the endometrial PMNs response after experimental challenge with *S. zooepidermicus* in mares and the effect of subsequent treatment with recombinant human IL-8. It was found that PMNs were attracted and translocated in the uterus within 6 h of administration of IL-8. This intra-uterine PMNs increased expression of the intra-cellular adhesion molecule of CD11α, in addition to an increased ability to generate reactive oxygen species.

vi. **Serum, plasma or hyper-immune serum:** Addition of a small amount of serum to uterine secretions can increase the opsonising capacity and enhance the phagocytic ability of PMNs significantly (Asbury, 1984). It is reported that 90% mares with subclinical endometritis and 52% with clinical endometritis became pregnant following i.u infusion of blood plasma (Ward, 1985). In another experiment on mare, Jochle (1998) reported that a series of uterine lavage with saline solution, followed by an instillation of 120 mL autologous plasma with or without oxytocin took 2.7 days to cure when the mares were experimentally infected with 1x109/mL *S. zooepidermicus* during estrus. In cattle, use of autologous plasma has been tried to cure endometritis. Venugopal (1995) observed 52.85% conception rate in endometritis cows treated with i.u infusion of homologous plasma. Methai (1999) observed 75% recovery rate and 50% overall conception rate after administrating autologous plasma (50 mL i.u at the dose rate of 1:10 of total bacterial load) and found a 70% overall conception rate.

vii. **PMN extracts and their components:** Bovine neutrophil extract has significant bactericidal activity against both gram-positive and negative bacteria (Gennaro *et al.* 1978). The extract of PMN cell contains three small peptides called defensins (Ganz *et al.* 1985). These defensins killed many bacteria and viruses *in-vitro* and could therefore, possibly be used as a therapy of bovine endometritis. In an *in-vitro* bacterial assay, Hughes and Couto (1988) showed that PMN extracts of rabbit origin had 100% killing of different bacterial isolates from equine uterus. They suggested that these PMNs could be utilized as non-antibiotic therapy for endometritis.

viii. Bacteria-free filtrate: Mycolic acid, which is a component of mycobacterial cell wall extracts , has been shown to cause a rapid and widespread influx of PMNs, which is chiefly responsible for the clearance of uterine infection, by stimulating TNFα, interferon-Ò, IL-6, IL-2 and myeloperoxidase production and by suppressing IL-10 production (Korf *et al.* 2005). Mycobacterial cell wall extracts was also used in endometritis mares that was experimentally induced with *S. zooepidermicus*, which resulted in the elimination of endometritis in 35% of the mares by the time of ovulation and 70% by 7 days post ovulation. They have also reported that mycobacterial cell wall extracts are effective in clearing endometrial infection by i.v or i.u administration (Rogan *et al.* 2010).

Effect of immunomodulators on reproductive hormones

In cattle, the effect of immunomodulators on the estrous cycle and peripheral blood hormone concentrations of inducing experimental endometritis appears to vary according to the infectious organism involved. After experimental uterine infection with *S. agalactia* on day 6 of the estrous cycle and infusion of lipopolysaccharide on day 9, the length of the luteal phase and blood progesterone concentrations of these cows were no different compared to controls (Hussain and Daniel, 1992). The i.u administration of lipopolysaccharide during the mid-cycle luteal phase had no effect on the length of the estrous cycle, or peripheral blood progesterone and prostaglandin F metabolite concentrations (Gilbert *et al.* 1990). A similar single treatment during estrus in heifers raised their serum cortisol concentrations; however, multiple infusions inhibited ovulation by suppressing the preovulatory surge of luteinizing hormone (Peter *et al.* 1989).

Using viable *E. coli* organisms to induce endometritis in heifers reduced the length of the estrous cycle (Gilbert *et al.* 1990). When *E. coli* endotoxin was administered using the i.u route to cows on day 5 after calving, peripheral blood concentrations of prostacyclin, PGF2α and thromboxane A2 metabolites were higher than in those treated on day 20 (Peter *et al.* 1990). The i.v administration of either *Salmonella typhimurium* or *E. coli* endotoxin in cows was associated with a transient increase in serum prostaglandin metabolites and a decrease in progesterone concentrations (Fredriksson, 1984; Peter *et al.* 1990). Generally, in new calved cows presenting signs of metritis or endometritis together with short cycles, plasma concentrations of the stable metabolites of prostacyclin,PGF2α and thromboxane A2 tend to be high (Peter *et al.* 1987). In summary, the inflamed endometrium permits absorption of endotoxin associated with natural infection or of exogenous origin, which increases peripheral blood concentrations of prostanoids.

Hence, in lieu of the different effects of immunomodulators, primarily simulating the infectious or inflammatory processes, has prevented their widespread adoption for treatment of uterine infections.

c. Enzymes

The use of proteolytic enzymes for the intramammary treatment of mastitis has been described by Zander (1997) and Kruger *et al.* (1999). Chymotrypsin, trypsin and papain have fibrinolytic and proteolytic activity in inflamed tissue; these properties have been utilized to treat infections. This effect is supposed to support the cellular defense mechanism. *In vitro*, enzymes inhibit the growth behavior and survival of micro-organisms such as *Staphylococcus* spp., *Streptoccocus* spp., *E. coli* (Kruger *et al.* 1999), *Arcanobacterium pyogenes*, some other aerobic bacteria and anaerobic bacteria, such as *Fusobacterium necrophorum* and *Bacteroides* spp. (Dohmen *et al.* 1995).

Chymotrypsin, trypsin and papain were infused i.u with 20g of Masti Veyxym1 (Veyx Pharma, Schwarzenborn, Germany), containing the enzymes chymotrypsin (16 mg), trypsin (16 mg), and papain (8 mg), respectively, and additionally 200,000 IU of retinolpalmitat (vitamin A) and 240 mg of α-tocopherolacetat (vitamin E) in postpartum cows at 21 to 27 days postpartum with vaginal discharge. A comparison with PGF2α was also made. Depending on the amount of pus in the vaginal discharge, the cows with vaginal discharge were categorized into 3 groups: mild endometritis (cows with clear mucus with flakes of pus), moderate endometritis (cows with mucopurulent discharge or fluctuating contents in the uterus) and severe endometritis (cows with purulent discharge with or without palpable contents in the uterus). The cows in different categories were randomly allocated to the two different treatments. While the enzymatic preparation was re-administered to the cows failing to show improvement, PGF2α was repeated in all the cows irrespective the response. Evaluation was made 14 days later. It was observed that the cure rate after the first treatment, defined as the absence of vaginal discharge at the re-examinations, was 59.7 and 68.0% in the cows administered enzymes and PGF2α, respectively ($P > 0.05$). Reproductive performance measures showed no significant differences between the two treatment groups. Service rate was significantly lower for enzymes and PGF2α, respectively, compared to control groups of cows. Conception rates to all services and %ages of cows pregnant by 250 days in milk were significantly lower in group administered enzymes compared to control, while no further differences were found between PGF2α and control. In both treatment groups, cure rate and reproductive performance measures were better for the cows with mild and moderate endometritis than for cows with severe endometritis. Conception rate to all services for cows with mild endometritis was significantly higher in cows treated with PGF2α than in group receiving enzymes. The results of this study suggested that PGF2α was superior to enzymes in the treatment of chronic endometritis in dairy cows (Drillich *et al.* 2005)

d. Hypertonic Dextrose

Sugar (hypertonic sucrose solution) heals wounds by inhibiting the growth of bacteria such as *E.coli* (Archer *et al.* 1990; Sharon, 2006). Some researchers believe that i.u infusion of 50% dextrose in cows with clinical endometritis improved uterine health through preventing local bacterial growth, increasing uterine tone and nurturing endometrial cells (Brick *et al.* 2012).

Accordingly, the i.u infusion of 50% dextrose solution (200 mL) in 28 to 35 days in milk, postpartum cows, with clinical endometritis was investigated. Except for the first service conception rate and proportion of cows pregnant before 100 days in milk, there existed no significant difference between the dextrose treated and the control group as far as reproductive performance is concerned. Approximately, 101 days postpartum, 50% of the healthy cows became pregnant, with the median days open in the dextrose group being 109. Hence, the use of a non antibiotic special solution of dextrose 50% is a good alternative to antibiotic agents concerning the treatment of chronic endometritis in dairy cows (Ahmadi *et al.* 2019).

e. Liquid Paraffin

Migration of peritoneal macrophages was observed following the intraperitoneal injection of liquid paraffin in mice, rats and rabbits (Hisatsune *et al.* 1983; Tsuru *et al.* 1984; Shacter *et al.* 1992; Inoue *et al.* 1997; Hayashi *et al.* 1999). Also, in a fire eater with pneumonia caused by the inhalation of liquid paraffin, macrophages and neutrophils were detected in bronchoalveolar lavage fluid (Yokohori *et al.* 2002; Burkhardt *et al.* 2003; Mylonaki *et al.* 2008). According to such reports, it seems that i.u infusion of liquid paraffin involves the migration of macrophages and other phagocytes into the uterus, thereby enhancing uterine defense mechanisms during infection (Lu *et al.* 2011). In a subsequent study Ahmadi *et al.* (2019) observed that cows with clinical endometritis receiving i.u infusion of 100 mL liquid paraffin compared to the healthy cows had similar median days to first AI. The reproductive performance was, however, significantly lower in cows receiving liquid paraffin. A lack in improvement in reproductive performance in the liquid paraffin infused cows can be due to an exaggerated response of uterus to produce neutrophils. In the study by Lu *et al.* (2011) the neutrophil numbers rose dramatically within 2 h of the infusion of liquid paraffin, the response being as similar to the cows receiving same volume of physiological saline. Moreover, the neutrophils reached peak levels approximately 8 h after the infusion in cows administered liquid paraffin and physiological saline. The peak level was maintained until 12 h after the infusion in both groups. Further, there were no differences in cytological reactions between the cows receiving either physiological saline or liquid paraffin during the initial 24 h after infusion and that the neutrophil level decreased relatively quickly to reach the pre-infusion level 2 days after the infusion of physiological saline. It is suggested that these acute responses of the uterus are not because of a liquid paraffin-specific reaction and instead may be a result of physical stimulus of the uterus by infusion of either liquid paraffin or physiological saline. Hence, both the ingredients can precipitate a non-specific defense reaction in the uterus. In contrast, the neutrophil level maintained its peak for 1 day after the infusion as alike saline group, but significantly higher numbers of neutrophils were detected in the liquid paraffin group compared with those in the physiological saline group on 3, 4, 5 and 8 days after the infusion followed by a gradual decrease through 8 to 10 days to reach the pre-infusion level 2 days before the subsequent ovulation in the liquid paraffin group. These findings indicate that continuous recruitment of neutrophils into the uterine lumen occurred in the liquid paraffin group. It was suspected that the persistence of liquid paraffin as a foreign body might stimulate the migration of phagocytes in the uterus of cows. Hence, the aforesaid limitations defy the logic of i.u infusion of liquid paraffin inspite of its augmented effect on migration of phagocytic cells.

f. N-acetylcysteine

Pyogenic bacteria, such as *Staphylococcus aureus*, *Escherichia coli* and *Pseudomonas aeruginosa* cause abscess formation and play a role in uterine infections. Importantly, *Klebsiella* spp. and *S. aureus* induce focal micro-abscess formation in the tissue (Smith and Risco, 2005; LeBlanc, 2008; Tras and Uney, 2011). Inflammation products throughout the infection are capable to alter pharmacokinetics and pharmacodynamics of antibacterial agents. Also, abscess formation leads to an acidic and anaerobic environment, therefore it is usual to see dormant bacteria and enzymes such as betalactamase propagating in the altered environment of infected uterus. In uterine infections, a uterine discharge including abscess material and tissue fragments is known to decrease the efficiency of aminoglycosides (Wagner *et al.* 2006).

N-acetylcysteine, a derivative of L-cysteine, is the precursor of glutathione. N-acetylcysteine reduces mucus viscosity by splitting the disulfide bonds linking proteins present in the mucoproteins and disintegrates the purulent or non-purulent secretions at the site of its presence. It is effective on fibrin and living tissue and its effect does not decrease in the presence of DNA. Hence, N-acetylcysteine is used as a mucolytic therapy and therefore in the treatment of obstructive pulmonary diseases, tuberculosis and cystic fibrosis in human medicine. In addition to the mucolytic effect of N-acetylcysteine, it also has antioxidant (Boothe, 2001, AHFS Drug Information, 2010), cytoprotective and anti-inflammatory effects (Kim *et al.* 2000; Olivos-Garcia *et al.* 2007; Tsai *et al.* 2010; Melkus *et al.* 2012). Moreover, it has been indicated that N-acetylcysteine has preventive and detractive effects on biofilm formation and bactericidal effects at the concentration of 80 mg/mL by deformation of uncolonized bacteria (Cammarato *et al.* 2012). N-acetylcysteine promotes the entrance of antibacterial agents into the mucus and its mucolytic effect does not depend upon the route of administration (Boothe, 2001). It has a good drug distribution potential all over the body including the uterus. Milk or meat withdrawal times of N-acetylcysteine following its administration as its legal withdrawal time according to the data of Europe-Middle East- Africa is 0 days (CVMP, 1996). Several authors stated that the i.u N-acetylcysteine administration in mares with endometritis may be beneficial in improving pregnancy and treatment rates and N-acetylcysteine has no adverse effects on the endometrial function (Gores-Lindholm *et al.* 2009; LeBlanc 2010; Melkus *et al.* 2012; Witte *et al.* 2012).

The effect of N-acetylcysteine on the treatment of clinical endometritis and pregnancy rate in dairy cows was investigated (Tras *et al.* 2014). The uterine discharge of the cows was ≥ 50% purulent. A group of 18 cows received an i.u infusion of amoxicillin trihydrate and potassium clavulanate (3500 mg and 875 mg) in 100 mL dosages after 12 h of an i.u infusion of 2% N-acetylcysteine during three days. In the non-N-acetylcysteine administered group, 100 mL of saline, instead of N-acetylcysteine, was administered in another group of 18 cows in the same protocol of the N-acetylcysteine -treated group. The clinical cure (83.3%) and pregnancy rates (66.7%) in the N-acetylcysteine-treated cows were significantly higher than those of non-N-acetylcysteine-treated group (55.5% and 27.8%, respectively). It was therefore concluded that the usage of N-acetylcysteine may increase the success of the antibiotic therapy in cows with clinical endometritis. A 0.3% solution of N-acetylcysteine had no influence on the oxidative burst of equine neutrophils, but a 3% solution accounted for the said damage (Le Blanc, 2012).

g. Estrogen

Though estrogen plays an important role in increasing the natural uterine defense mechanism through increasing the blood circulation, leucocytic infiltration into the uterus, uterine contraction and mucus flow (Roberts, 1986), it is currently seldom used in the treatment of endometritis in lactating cows because of its considerable levels that have been detected in milk that may make the milk of the treated cows unfit for human consumption (Roberts, 1986). Furthermore, the use of estrogen in higher doses could lead to the development of ovarian cysts (Arthur *et al.* 2009).

h. Ozone

Ozone is a gaseous molecule comprising of three oxygen atoms (O3). It is created naturally in the atmosphere by lightning during thunderstorms. It rapidly degrades back to oxygen atoms at room temperature. Ozone has been utilized in bovine medicine as a treatment of endometritis when applied into the uterus as a foaming spray through a catheter. Ozone has germicidal and fungicidal action by virtue of having a very powerful oxidant property and is known to react very quickly. Ozone breaks through the microorganism (bacteria and germs) cell membrane, and also destroys viruses by diffusing through the protein coat in the nucleic acid core, resulting in damage of the viral nucleic acid. The advantage of ozone use is how it blocks microorganism resistance in contrast to antibiotics which may generate resistance (Scrollavezza *et al.* 1997). First applications of ozone in medicine were limited to contact with the skin, as disinfectant, and then oxygen–ozone therapy as direct injection of a gaseous mix with molecular oxygen. Germicidal and fungicidal action of ozone is provoked by the oxidative process caused by ozone and peroxides that destroyed the capsule.

There are many different forms of ozone preparations (cream, gas, injections, pellets, foam, pearls, boluses). The use of ozone in veterinary medicine is in the form of foam and pearls, which have been used to treat endometritis and metritis (Scrollavezza *et al.* 1997; Marusi *et al.* 1999; 2000; Đurièiæ *et al.* 2012). The use of ozone oil derivatives can be advantageous from a biochemical point of view in the treatment of several pathologies. The most important derivatives are lipohydroperoxides, which act directly on pathogens or stimulate the cellular defense systems. Major uses of lipohydroperoxides in veterinary medicine are in the form of foam and ovuli for the treatment of endometritis and metritis (Scrollavezza *et al.* 1997; Marusi *et al.* 1999, 2000). Moreover, the use of ozone's lipohydroperoxide products does not produce any residue in food such as meat and milk.

The i.u ozone treatment alone or combined with parenteral antibiotics was a more efficacious treatment for retained placenta in cows when compared to hormonal and parenteral antibiotic treatment modalities (Ðurièiæ *et al.* 2012; Zobel and Tkalèiæ, 2013). Cows with purulent endometritis need to be re-administered the ozone treatment (according to the manufacturer's recommendation) at least twice in a three week period in accordance with the estrous cycle.

The success of ozone therapy was determined in cows suffering from metritis diagnosed on days 5 and/or 15 following parturition as well as in another set of cows diagnosed with endometritis on days 25 and/or 45 after parturition. The third group included animals without clinical signs of uterine inflammation. In cases when metritis or endometritis was diagnosed, ozone foam (Riger spray G) was applied into the uterus. Ozone was applied by means of a vial containing ozone foam under pressure for 5 seconds. The time to the first postpartal insemination was significantly shorter for controls compared to the metritis and endometritis groups. The days open until pregnancy was the longest (133 days) in cows with endometritis. In the control group, the days open until pregnancy was 125 days and in the group of cows treated for metritis was 120 days. Services per conception for the cows with metritis, endometritis and control cows were 1.86, 2.21 and 1.90, respectively. The i.u ozone flush therefore exhibited potential to alleviate metritis and endometritis as an efficacious and cost=effective treatment option with an overall positive effect on fertility (Djuricic *et al.* 2012).

In another study, the influence of preventive application of two different ozone preparations on reproductive efficiency in dairy cows with physiological puerperium was investigated in Holstein cows. The first group was treated using 6 i.u ozone pearls (RIPROMED OVULI O3®, Novagen, Italy) once, whereas the second group consisted of cows treated with ozone foam spray (RIGER SPRAY , Novagen, Italy) application in the body of uterus for 5 seconds. The treatments were made 24 to 48 h after parturition. It was observed that preventive ozone i.u application during early puerperal period, especially the foaming form of ozone preparations (spray), improved the reproductive efficacy in dairy cows by 17 and 12 days, respectively (Djuricic *et al.* 2012). In a subsequent study involving cows at day 35 after calving, instillation of 50 mL of ozonized intrauterine distilled sterile solution (50 ìg/mL) in the uterus reduced subclinical endometritis prevalence and improved reproductive performance in postpartum dairy cows (Escandon *et al.* 2020).

i. Protein, VPI-O22

A natural, novel protein, VPI-O22 that could inhibit bacterial infection and restore uterine tone and health was evaluated in cows with clinical endometritis. Two hundred mL of VPI-O22 (5%) were infused i.u and the treated animals were assessed 7 days later. Rectal temperature was higher in cows with endometritis and was found to be normal after one week of VPI-O22 treatment. Mean pH of uterine discharge from metritis cows was 8.8 (control 6.8) and changed to 7.0 after one week of VPI-O22 treatment. Mucus score was 6 times higher, and total number of PMN cells was more than 70 times higher in cows with metritis and significantly reduced (2 and 4 times higher, respectively, compared with control) after VPI-O22 treatment. Bacterial culture of uterine swab demonstrated the presence of different types of bacteria including gram positive, gram negative, α- and β-hemolytic, rods, and cocci. Zone of inhibition studies with different concentrations of VPI-O22 demonstrated that 5% would be more than enough to kill all types of bacteria found in these studies. Hence, it was concluded that VPI-O22 is a potent antimicrobial protein and has the ability to kill different types of bacteria in *in-vitro* and clinical applications (Talukdar *et al.* 2018)

j. Nonsteroidal Anti-inflammatory Drugs (NSAIDS)

The use of NSAID has been studied as a supportive treatment (Drillich *et al.* 2007) or as the first treatment option (Pohl *et al.* 2016) for metritis. Although NSAID did not seem to improve reproductive efficiency or milk production when used in combination with systemic antimicrobials (Drillich *et al.* 2007), they reduced the number of animals that required systemic antimicrobials when used as the first treatment choice (Pohl *et al.* 2016). However, the most important benefit of using NSAID could be to improve animal well-being during a painful and stressful disease (Laven *et al.* 2012). The latter study found that 42% of the cow dairies used NSAIDs as a supportive treatment for metritis. In the United States, flunixin meglumine has a 36-h milk withhold and 4-d meat withdrawal period; this probably discourages its use. Currently, no FDA-approved drugs exist for pain management in adult dairy cattle, and all current use of drugs for this purpose occur extra-label.

It is known that as long as the PGF2α release dominates, as is observed during postpartum uterine inflammation, ovulation does not occur. When the PG metabolite levels have returned to baseline, ovulation can occur (Kindahl *et al.* 1984). Lindell *et al.* (1982) and Madej *et al.* (1984), however, showed a significant negative correlation between the duration of PG release and uterine involution. In the study of Königsson *et al.* (2002), an early (3 to 6 days postpartum) flunixin treatment, when the release of PG is high, also suppressed levels of PG metabolites during the treatment period of cows with postpartum uterine inflammation. But this suppression was not complete, and not so severe, as in case of administration of flunixin at the end of the second week postpartum. Guiltbault *et al.* (1987), in contrast, did not detect a negative effect of flunixin meglumine on uterine involution. Findings similar to the latter were also echoed by another set of workers (Jeremejeva *et al.* 2012).

There are few other limitations from use of flunixin. Flunixin did not improve the general depression in the cows showing fever and reduced appetite and may be disadvantageous as there is an increased risk for rectal bleeding in flunixin treated cows. Moreover, early flunixin treatment during a period of acute disease and stress might contribute to an increased sensitivity to adverse effects (Konigsson *et al.* 2001). Odensvik and Fredriksson (1993) reported negative gastrointestinal side-effects of flunixin treatment in their study, where the intensive medication was suspected of causing diarrhea and deprived appetite in the treated animals. Furthermore, treatment with COX inhibitors, such as flunixin, is aimed at diminishing the inflammatory reaction. This might be hazardous in infectious diseases as it might push the host-parasite balance in a negative direction, leading to a more severe infection. In the latter context, it needs to be understood that after experimental endotoxin administrations animals become refractory to subsequent endotoxin administrations (Bieniek *et al.* 1998). The release of endotoxin during a natural infection gradually increases over time, in parallel with the progression of infection. Hence, over the course of infection (experimental or natural), systemic mechanisms mediating the refractoriness to endotoxin might be activated during an early phase of the infection. Under such circumstances flunixin administration may not improve the clinical outcome (Konigsson *et al.* 2001).

k. Phytotherapy

The use of herbal remedies for the treatment of livestock dates back to 5000 BC as depicted in Nakula Samhita, but this traditional system of medication has picked up importance in the recent years (Mandhwani *et al.* 2017). The following plant based products have been used in treatment of uterine infections.

Garlic (*Allium sativum*): Historically, garlic has been used for centuries worldwide by various societies to combat infectious disease. Louis Pasteur was the first to describe the antibacterial effect of onion and garlic juices. One of the active principles of freshly cut garlic homogenates is allicin, which has a variety of antimicrobial activities, and the antibiotic activity of 1mg of allicin is equated to that of 15 IU of penicillin. Garlic can be provided in the form of capsules and powders as dietary supplements and thus differs from conventional foods or food ingredients (Sadanandan *et al.* 2014).

Sarkar *et al.* (2006) evaluated the efficacy of gar-lic extract and PGF2α in the treatment of endometritis in cows. After treatment, there was a significant reduction in bacterial load, whereas it increased in control group. The estrual cervicovaginal mucus turned clear in 70% animals treated with garlic extract. The overall conception rate was 50% in the treated groups as compared to nil pregnancy in the control group. In a subsequent study, Rahi (2011) found 50% conception rate and 75% recovery rate of endometritis when treated with aque-ous extract of garlic.

Neem (*Azadirachta indica*): Neem is the most commonly used traditional medicinal plant in India. Almost all parts of the neem plant are endowed with medicinal properties, and it has been demonstrated to exhibit immunomodulatory, anti-inflammatory, antifungal, antibacterial, antiviral, and antioxidant properties. *Azadirachtin*, found only in *Azadirachta* spp. is a complex tetranortriterpenoid. Of all of the liminoids in neem, azadirachtin and its 25 natural analogs are biologically most active (Devakumar and Kumar, 2008).

Kumar *et al.* (2013) reported that the hydroalco-holic and hydroacetonic extracts of the neem have potent immune-modulatory and therapeutic efficacy on endometritis in repeat breeding crossbred cows. In their study, a significant rise was found in total leu-kocyte count, PMN cells and immunoglobulin con-centration in the treated cows.

Tulsi (*Ocimum sanctum*): *O. sanctum* L., known as "tulsi" in Hindi and "holy basil" in English, is an erect softy hairy aro-matic herb or under-shrub found throughout India. *O. sanctum* L. is held sacred by Hindus and is used as a medicinal plant in day-to-day practice for vari-ous ailments. In Ayurveda, tulsi has been well docu-mented for its therapeutic potentials. Different parts of tulsi plant, for example, leaves, flowers, stem, root, and seeds, are known to possess therapeutic potentials and have been used by traditional medical practitioners as expectorant, analgesic, anticancer, antiasthmatic, antiemetic, diaphoretic, antidiabetic, antifertility, hepatoprotective, hypotensive, hypolip-idemic, antimicrobial, and antifungal activity against *Aspergillus niger* (Gupta *et al.* 2002).

Eugenol is a phenolic compound and major con-stituent of essential oil extracted from different parts of tulsi plant. Eugenol is the most prominent phyto-constituent present in the tulsi plant, which may be responsible for antimicrobial activity (Bhatt *et al.* 2012).

Ashwagandha (*Withania somnifera*): Ashwagandha is an important herb in the ayurvedic and indigenous medicinal system for over 3000 years. It contains potentially active constituents such as Sitoindosides VII-X and Withaferin A. It is a sedative, diuretic, anti-inflammatory agent, and is generally used for increasing energy and endurance, and acts as an adaptogen that exerts a strong immune-stimulatory and anti-stress effects (Umadevi *et al.* 2012).

Owis *et al.* (2005) found the antibacterial activity of alcoholic extract made by ashwagandha roots against *Salmonella typhimurium*. Moreover, the extract did not induce lysis on incubation with human erythro-cytes advocating their safety to the living cells.

Turmeric (*Curcuma longa*): Turmeric is an ancient spice derived from the rhizomes of *C. longa* and is also known as "golden spice of India." It has been found that both curcumin and the oil fraction suppress the growth of several bacteria such as *Streptococcus* spp., *Staphylococcus* spp. and *Lactobacillus* spp. (Rathaur *et al.* 2012).

***Viburnum opulus* (Guelder-rose):** It is the species of flowering plant, family *Adoxaceae* (formerly *Caprifoliaceae*) native to Europe, northern Africa and central Asia. The bark, known as cramp bark, is employed in herbal medicine. Its use was formerly included in the United States Pharmacopoeia, but is now omitted though it has been introduced into the National Formulary in the form of a fluid extract, compound tincture and compound elixir, for use as a nerve sedative and anti-spasmodic in asthma and hysteria. *V. opulus* is a powerful antispasmodic and is much used in the treatment of asthma, cramps and other conditions such as colic or painful menstruation. It is also used as a sedative remedy for nervous conditions. The bark is antispasmodic, astringent and sedative. The bark contains "scopoletin", a coumarin that has a sedative affect on the uterus. Its use as tea relieves all types of spasms, including menstrual cramps, spasms after childbirth and threatened miscarriage. An endometritis related study was conducted as induced by suturing 15mm piece of endometrium into abdominal wall of Sprague Dawley rats and the effectiveness of the fruit extract of *V. opulus* was found to be beneficial because of chlorogenic acid. Other phenolic compounds could potentiate the activity due to their amount (Sharma *et al.* 2018).

Other plants based immunomodulatory studies: Borowieck (1989) observed about 84% conception rate in endometritis cows treated with alcoholic extracts of 5 plants (chamomyle, marigold, comfrey, salvia and yarrow) for endometritis.

Phytotherapy in combination: Owis *et al.* (2005) showed that ashwagandha and garlic are antibacterial and immune-modulatory in nature and thus reduce bacte-rial load and subsequently inflammation process.

Kumar (2013) found better therapeutic effects of garlic, neem, and garlic in combination with turmeric and neem. This was evident by observing significant increases in total leukocyte count and PMN% in uterine flushing after treatment with herbs. Garlic alongwith ashwagandha extract was found as the most effective treatment among all treated groups and, thus, can replace conventional antibiotic in future for bacterial endometritis leading to repeat breeding con-dition in crossbred cows (Rahi *et al.* 2013).

Singh (2016) found that the cows exhibited clear discharge with ciprofloxacin (93%) followed by gar-lic (88.3%) and neem (86.7%) after treatment; and found a significant difference in PMN%, total leukocyte, lymphocyte and monocyte counts.

I. Silver Nanoparticles

Nanoparticles appear to be suitable and alternative antimicrobial agents to overcome the alarming rate of the spread of antibiotics resistance, toward improved detection and killing of pathogenic bacteria. Recently, several studies have demonstrated plant and microbial extracts, essential oils, pure secondary metabolites, and newly synthesized molecules as potential antimicrobial agents (Runyoro *et al.* 2006; Mabona *et al.* 2013; Nazzaro *et al.* 2013).The application of nanoparticles has attracted huge interest in several fields, including biotechnology, biomedical sciences, and veterinary medicine. Several studies have explored the possibility of high-level nanotherapy in humans. However, the applications of nanotechnology in veterinary medicine have not reached the same level, and remain in a relatively innovative stage. Of late, nanoparticles have been used as nutraceuticals, biocides, diagnostic tools, reproductive aids, and in drug and nutrient delivery in veterinary medicine (Hill and Li, 2017), and show potential to serve as alternatives to conventional antimicrobial agents (Bengtsson and Greko, 2014; Yah and Samate, 2015). In this regard, nanoparticle-based therapy is a promising approach to improve the balance between the efficacy and toxicity of systemic therapeutic intervention.

Among the various metal nanoparticles available, silver nanoparticles (AgNPs) have attracted tremendous interest in biomedical applications, including for antimicrobial therapy, wound dressings, diagnosis and treatment, and contraceptive devices (Naidu Krishna *et al.* 2015). Moreover, AgNPs have been used as sensors, imaging, drug delivery, and for tissue engineering in veterinary medicine and animal production (Narducci, 2007). Thus, AgNPs represent a very promising therapeutic agent with unique potential against various microbial pathogens, with a particularly high capacity to effectively act on antibiotic-resistant bacteria (Leid *et al.* 2012; Yah and Simate, 2015). To date, AgNPs have been widely used as effective antimicrobial agents against various bacteria, fungi, and viruses (Gurunathan *et al.* 2014a; Zhang *et al.* 2016). AgNPs can potentially inhibit multiple drug-resistant strains of *Staphylococcus aureus* and *Pseudomonas aeruginosa* that cause mastitis (Leid *et al.* 2012), and have proven to be effective against various antibiotic-resistant bacteria (Gurunathan *et al.* 2014a; Gurunathan, 2015; Zhang *et al.* 2016). The mechanisms of inhibitory action of AgNPs are attributed to their high reactivity with bacterial proteins, sugars, and DNA, resulting in structural alterations to the cell wall and the membrane, ultimately leading to inhibition and cell death (Yuan *et al.* 2017).

The AgNPs were synthesized and characterized using the biomolecule apigenin as a reducing and stabilizing agent. The AgNPs so produced was evaluated against multiple drug-resistant gram-negative and gram-positive bacteria, including *Prevotella melaninogenica* and *Arcanobacterium pyogenes* that were isolated and characterized as predominant isolates from clinical endometritis samples. The mechanism of toxicity of AgNPs in *P. melaninogenica* and *A. pyogenes* was also evaluated. The findings suggested that AgNPs inhibit bacteria in a dose-dependent manner, and the minimum inhibitory concentration values of AgNPs against *P. melaninogenica* and *A. pyogenes* were found to be 0.8 and 1.0μg/mL, respectively. The bactericidal effect of the AgNPs was dependent on several factors, such as the concentration of AgNPs, size, shape, physicochemical properties, and the initial bacterial concentration. In general, AgNPs showed better antimicrobial activity against the gram-negative bacterium *P. melaninogenica* when compared to that against the gram-positive bacterium *A. pyogenes*, which has also been echoed in some similar previous reports (Gurunathan *et al.* 2014b; 2015). Also, AgNPs inhibited cell viability and biofilm formation in a dose- and time-dependent manner. Silver ions are responsible for formation of free radicals. The metabolic toxicity of the AgNPs was caused by an increase in oxidative stress, as evidenced by the increased generation of reactive oxygen species, malondialdehyde, protein carbonyl content, and nitric oxide. It was concluded that biomolecule-mediated synthesis of AgNPs shows potential as an alternative antimicrobial therapy for bovine metritis and endometritis (Gurunathan *et al.* 2018). In a subsequent study (Hussein and Hussein, 2022), 20 mL (100 mg; @0.2 mg/kg body weight) i.u infusion of silver nanoparticles was undertaken to treat clinical endometritis in Iraqi cows. Green biosynthesis was used to synthesize silver nanoparticles from *Curcuma longa* tuber powder. The prepared Ag/*C. longa* was characterized by X-ray diffraction, Fourier-transform infrared spectroscopy and scanning electron microscopy (Shameli *et al.* 2012). A cure rate of 70% (20% in the untreated control) as assessed on the basis of color and pH of uterine mucus, endometrium thickness, percentage of PMN cells in the uterine mucus, was reported.

m. Dimethyl Sulfoxide, Hypochlorous Acid, Ozone and Anitmicrobial Peptides in Relation Biofilm Coated Microbes

Bacteria are capable of living in two different lifestyles: planktonic or biofilm states. Distinct bacterial communities are able to thrive through virulence mechanisms such as mucin degradation, biofilm formation, and antimicrobial resistance, leading to dysbiotic states (Yeoman *et al.* 2010; Ferris, 2021). A biofilm is a community of bacteria that are attached to an interface or to each other that are encased within an extracellular polymeric substance consisting of nucleic acids, lipids, proteins, and exopolysaccharides. Bacteria residing in a biofilm exhibit altered phenotypes with respect to growth rate and gene transcription as compared to planktonic bacteria (Chen *et al.* 2017). The production of an extrapolymeric substance, which is key components of the biofilm matrix, results in reduced host recognition and antibiotic tolerance. Biofilm development is likely to contribute to chronic or recurrent uterine infections in equines. Although there is no consensus on the endometrial microbiota, commensal bacteria inside the uterine cavity could also help maintain a eubiotic state. In humans, bacterial colonization of the endometrial epithelia through biofilms were reported in the endometrium from patients diagnosed with bacterial vaginosis provoked by *G. vaginalis* (Swidsinski *et al.* 2013; Franasiak *et al.* 2016). Biofilm infections are often multifocal, with the greatest amount of adherent material located between the endometrial tissue folds and in the uterine horns. Culture and cytology alone are unable to differentiate a biofilm and planktonic-associated bacterial infection.

Staphylococcus epidermidis, Staphylococcus aureus, Pseudomonas aeruginosa, Klebseilla pneumonia, Streptococcus pneumonia, E. coli, Helicobacter pylori, yeasts and fungi cause a biofilm formation, which may impair the drug exposure, contribute to chronic infection and drug resistance. Some antibacterial agents such as macrolides and clindamycin are not capable of treating the infection against the bacteria producing biofilm, unlike these agents have more efficiency than other bacterial agents (Le Blanc and Causey, 2009; Tras and Uney, 2011). It has been reported that the combination of antibiofilm and antibacterial agents may induce the drug synergism due to the increasing sensitivity of biofilm producing bacteria to antibacterial agent by deformation of biofilm (Cammarato *et al.* 2012). Clinically, a biofilm may be suspected if a uterine infection is treated appropriately, including antibiotic therapy based on antimicrobial susceptibility tests, and yet the infection persists or reoccurs.

In vitro studies indicated that a combination of antibiotics and non-antibiotic agents, such as Tris-EDTA or dimethyl sulfoxide, were more effective against a bacterial biofilm than treatment with either an antibiotic or a non-antibiotic agent alone.

Dimethyl sulfoxide is a solvent with anti-inflammatory properties that has been used clinically in mares to eliminate mucus, exudate, or biofilms (LeBlanc, 2010). Ulceration of the epithelium and invasion of inflammatory cells was associated with higher concentrations (75% [vol/vol]) of Dimethyl sulfoxide (Frazer *et al.* 1988). The endometrium instead exhibited no signs of ulceration or inflammatory cells 21 days after infusion of the 30% solution of dimethyl sulfoxide (Zobel *et al.* 2012).

Hypochlorous acid is a bactericidal component of the innate immune system produced by activated neutrophils. Hypochlorous acid has been reported to be bactericidal by inducing a rapid decrease in ATP production and disrupting the integrity of the bacterial cell membrane (Barrette *et al.* 2002; Wang *et al.* 2007). Hypochlorous acid (0.35 to 0.7 mg/mL) has antimicrobial activity against *E. coli*, *P. aeruginosa*, and *K. pneumoniae* (Wang *et al.* 2007). Because infection delays wound healing, topical hypochlorous acid has been used as wound cleansing agents. Currently, there are several hypochlorous acid agents marketed in veterinary medicine for wound care, ophthalmic ointment and wash, ear wash, and naval dip. In the study of Loncar *et al.* (2017) neither of the two hypochlorous agents tested were able to reliably disrupt a preformed biofilm or eliminate bacteria. This could have been due to the inherent instability of hypochlorous acid, which could compromise *in vivo* treatment (Wang *et al.* 2007).

Ozone has demonstrated efficacy when used to treat oral biofilms (Magni *et al.* 2008) and in food and water sanitation (Güzel-Seydim *et al.* 2004). Ozone is a potent oxidizing agent that can attack bacterial membrane glycoproteins and/or glycolipids resulting in cell disruption and leakage of cell contents (Güzel-Seydim *et al.* 2004). It has demonstrated strong bactericidal effects, especially against gram-positive organisms (Duric *et al.* 2015).

Antimicrobial peptides are part of the innate immune system. These molecules are able to interact with and disrupt bacterial plasma membranes. Once bound, antimicrobial peptides induce a conformational change and transverse the lipid bilayer, producing pores to form in the membrane that results in the leakage of metabolites and ultimately cell death (Yeaman and Yount, 2003). Unfortunately, some organisms have demonstrated both passive and adaptive resistance mechanisms to antimicrobial peptides (Yeaman and Yount, 2003). Antimicrobial peptide mimics have been developed which show increased resistance to proteolysis and are more cost-effective to produce on a large scale. One such mimic has been developed from ceraginins, a family of bile acid derivatives. CSA-13 is the most potent of the ceraginins developed and has been reported to be efficacious against a *P. aeruginosa* biofilm *in vitro* (Epand *et al.* 2010). In a 2010 study,

CSA-13 was shown to have bactericidal activity against mature *P. aeruginosa* biofilms formed *in vitro* (Nagant *et al.* 2010)

In one study, the *in vitro* effectiveness of dimethyl sulfoxide, hypochlorous acid, ozone, or an antimicrobial peptide mimic at disrupting a preformed biofilm was made and the bacterial killing of different gram-negative isolates from the equine uterus was also evaluated. Dimethyl sulfoxide in the acute puerperal metritis reliably eliminated preformed biofilm and reduced colony forming units of *Escherichia coli* and *Klebsiella pneumonia*, but were only effective against 10% to 75% of the *Pseudomonas aeruginosa* isolates. The minimum biofilm eradication concentration for dimethyl sulfoxide at disrupting an *E. coli* biofilm was 30% vol/vol, and the concentration for dimethyl sulfoxide to reduce CFUs was 15% vol/vol for *K. pneumonia*. The minimum biofilm eradication concentration for antimicrobial peptides for reducing biofilm mass was 0.5% of the as directed solution for *E.coli* and *K. pneumoniae*. The minimum biofilm eradication concentration of antimicrobial peptide for reducing CFUs was 0.5% of the commercial solution for *K. pneumoniae*. Ozone and the two hypochlorous agents were not reliably able to reduce biofilm or kill any of the gram-negative organisms tested. A lack of effect of hypochlorous acid on biofilm could have been due to the inherent instability of hypochlorous acid, which could compromise *in vivo* treatment (Loncar *et al.* 2017). These bacterial species-specific results stress the importance of acquiring organism identification when developing a treatment plan for bacterial endometritis because not all treatments are effective against every bacterial species. Interestingly, some NSAIDs, such as diclofenac, ibuprofen, and particularly aspirin, have antimicrobial properties by inhibiting processes such as bacterial growth, adhesion, and biofilm formation (Medina-Bastidas *et al.* 2022).

n. Peroxyethane Acid

Peroxyethane acid is capable of disrupting the ring of bacterial catalase and has a broad microbiocidal spectrum. Following application, peroxyethane acid degrades to acetic acid and peroxide. There are no residues and no resistance development. Because peroxyethane acid administration for bladder and wound lavage in human medicine produced positive results, the substance was tested for the treatment of cows with endometritis.

After the influence of organic acids or peroxide acids with low concentration upon mucous membranes results a higher activity of the topical defense-situation. The application of a 0.2 per cent solution of peroxyethane acid (drug Uterofertil) into the uterus for endometritis-therapy directs an increase of phagocytosis-activity not only topical but also in the peripheral circulation of blood. The phagocytosis-activity was measured with a test by granulocytes (Strube *et al.* 1991).

The effect of 0.2% solution of peroxyethane acid was evaluated and compared with antibiotics in cows with chronic catarrhal endometritis. After peroxyethane acid use, equivalent or better results were achieved compared to antibiotic-treated cows. Histological and cytological examinations demonstrated a rapid and high regenerative ability of the endometrium after i.u peroxyethane acid instillation. Peroxyethane acid induced an intensive phagocytosis beyond the 6th day of instillation. Hence, peroxyethane acid may be used as an alternative to antibiotics to treat endometritis. Currently, no licensed product is available, which significantly restricts peroxyethane acid use (Busch, 2018).

o. Moist Heat Steam

"Samia-treat" (SAT), has been another approach in which repeat breeder cows with subclinical endometritis were given a local i.u moist heat treatment by infusing 50 mL sterile boiling water (H"100°C) once. The findings were compared with other groups of cows administered i.u with antibiotics comprising of either 50 mL of 5% oxytetracycline for three successive weeks or 10 mL of 10% gentamicin along with 10 mL of penicillin and streptomycin in 30 mL saline for three successive weeks. Leukogram revealed neutrophilia in antibiotic treated groups 36 h and 1 week after treatment, whereas it reduced on the two occasions in the cows treated with boiling water. The pregnancy rate to first three inseminations was 83.3% in the cows treated with boiling water compared to 40% and 50% in the cows given oxytetracycline and gentamicin with streptopenicillin, respectively. The number of animals in each group ranged from 10 to 12, which are considered to be low for recommendation / use of the proposed treatment (El-Rheem *et al.* 2019).

E. Comparison of Treatments

a. *E.coli* Lipopolysaccharide *Versus* Oyster Glycogen *Versus* Enrofloxacin

In a comparative study involving treatment of endometritis cows using either of 100 ìg of *E. coli* lipopolysaccharide, 500 mg oyster glycogen, both infused once, and enrofloxacin, which was infused i.u for 3 consecutive days, were evaluated. Endometrial biopsy was collected at 0 h and 72 h from all the cows for histopathological study. Uterine flushing was collected at estrus prior to infusion of the drugs (0 h) and at 24, 48 and 72 h post treatment for evaluation of changes in total protein concentration, total leucocytic count, PMN cell %, and the total immunoglobulin concentration. It was found that the *E. coli* lipolysaccharide and oyster glycogen treatment significantly increased the total protein concentration, leucocytes, PMN cell % and the concentration of total immunoglobulin at 24, 48 and 72 h, post treatment. However, i.u enrofloxacin therapy reduced the total protein concentration and PMN cell% at 72 h post treatment than pretreatment levels. The recovery rate was higher both in lipoploysaccharide and oyster glycogen groups (83.3%) than the control group (25.0%). But in antibiotic group, 75.0% of the treated cows showed clinical recovery from endometritis as against the control. The conception rate was significantly higher with lipoploysaccharide and oyster glycogen (75%) as well as with enrofloxacin (58.3%) treatment compared to the control (16.7%). Hence, *E. coli* lipoploysaccharide and oyster glycogen may show better therapeutic efficacy to stimulate non-specific uterine defense mechanisms in crossbred cows suffering from endometritis (Prasad *et al.* 2009)

b. Hypertonic Dextrose *Versus* Liquid Paraffin *Versus* Oxytetracycline *Versus* PGF2α

Cows with clinical endometritis diagnosed at 28 to 35 days in milk, were evaluated for their reproductive performance following treatment with an i.u infusion of 200 mL of 50% dextrose solution, 100 mL of liquid paraffin, a single PGF2α injection or i.u infusion of 5 g of oxytetracycline. The incidence of endometritis was 41.6% in this study. Based on the results of different reproductive indices, the reproductive performance was significantly lower in liquid paraffin group compared with the healthy cows and other treatment groups. Except for the first service conception rate and proportion of cows pregnant before 100 days in milk, there existed no significant difference between the dextrose group and the control. The first service conception rate was lower in dextrose compared to oxytetracycline group. It was concluded that the use of 50% dextrose is a good alternative to antibiotic agents for the treatment of clinical endometritis in dairy cows (Ahmadi *et al.* 2019). Reduced effectiveness of liquid paraffin compared to the other agents on the reproduction indices can be attributed to an exaggerated inflammatory response which might stimulate an extended migration of phagocytes in the uterus of cows.

c. Ceftiofur *Versus* Two Doses of Cloprostenol (at 14-d interval)

The systemic antibiotic treatment of clinical endometritis in dairy cows (21 to 27 days in milk) with ceftiofur (1 mg/kg; i.m; 3 consecutive days) was compared with a treatment protocol consisting of two doses of PG (cloprostenol; 0.5 mg; at 14-d interval). The cows were re-examined at 42 to 48 d post calving. Proportion of cows cured, i.e. cows with clear, translucent or no mucus (74.2 and 80.2%)

was similar for two treatments. Also, the voluntary waiting period, set at 40 days in milk, AI submission rate, days to first service, first service conception rate, days open and proportion of cows pregnant did not differ between the groups. The cows were yet to resume cyclicity at the time of either treatment (Kaufmann *et al.* 2010b). In the absence of corpus luteum, the use of PGF2α for the treatment of endometritis is efficient and might work by induction of myometrial contractions and estrus (Weems *et al.* 2006), as also by regulation of inflammatory responses in the endometrium. It has been demonstrated that bacterial endotoxin lipopolysaccharide, expressed by uterine pathogens such as *E. coli* switch the endometrial epithelial secretion of prostaglandins from the F to the E series. This results in prolonged luteal phases and modulation of endocrine functions that are essential for reproduction (Herath *et al.* 2009; Sheldon *et al.* 2009b). Therefore, the exogenous administration of PGF2α might compensate the reduced endogenous epithelial secretion of PGF2α. In contrast, other authors reported that lipopolysaccharide stimulates the secretion of PGF2α (Leung *et al.* 2001). Results by Gabler *et al.* (2009) suggested an association between a dysregulated cytokine and/or prostaglandin profile in the endometrium and the occurrence of clinical and subclinical endometritis.

In another study, a combined treatment with PGF2α and ceftiofur showed an even stronger treatment effect than PGF2α alone (Haimerl *et al.* 2017)

d. PGF2α *versus* Cephapirin

The time to pregnancy in dairy cows diagnosed with clinical endometritis between 20 to 23 days in milk revealed no significant difference in days open after treatment with i.u infusion of cephapirin benzathine and those injected with PG (LeBlanc *et al.* 2002b). The time of injection, ovarian structures and the dosage of PG are important factors that may enhance productive performance in cows with endometritis (LeBlanc, 2008). For instance, a high proportion of cows have spontaneous resolution of endometritis until at least 4 wk postpartum and a therapeutic intervention thereafter may be more successful. Irrespective of the type of treatment (PGF2α or cephapirin), the response was better in cows that had re-established the ovarian cyclicity; the response to antibiotic therapy was, nevertheless better than PGF2α. Hence, therapeutic intervention should be reserved for cases diagnosed after 26 days postpartum.

e. PGF2α *Versus E. coli* Lipopolysaccharide

Using a small number of cows, early puerperal uterine immune stimulation by double systemic cloprostenol injections (8 and 9 days postpartum) or single i.u *E.*

coli lipopolysaccharide infusion (day 7 postpartum) did not result in significant protection from endometritis or improvement of open periods. However, there is indication as to prophylactic potentials of early puerperal i.u *E. coli* lipopolysaccharide treatment, at least in terms of estrus detection prior to 120 days postpartum, compared to the untreated group. However, variation in the cows at the treatment onset prevents the usefulness of the findings *per se* (Wakayo *et al.* 2015).

f. PGF2α and Cephapirin *Versus* 50% Dextrose Solution *Versus* Liquid Paraffin

There was no difference in terms of cytological and clinical cure rate between the treatments (Makki *et al.* 2017).

g. PGF2α or NSAID in Combination with Antibiotics

In cows with acute puerperal metritis and clinical metritis, treatment was started on the 3rd day postpartum using administration of ceftiofur (s.c. injection of 1 mg/kg) for five days in combination with flunixin (1.1 mg/kg) for first three days. Another group received ceftiofur (1 mg/kg) for five days followed by two injections of PGF2α, with an interval of 8 h, on the 8th day postpartum. No differences were found for vaginal discharge. There were no differences in bacterial growth, start of ovarian activity or serum amyloid-A or fibrinogen levels among the groups. The haptoglobin concentration was higher in the first and second weeks postpartum in group receiving antibiotic and prostaglandins compared with the other groups including untreated control. The number of days open was higher in group receiving antibiotic and flunixin than in other two groups. A higher pregnancy rate and lower number of services per conception after the first two services was higher in ceftiofur combined with PG groups, which inspite of more severe uterine inflammation prior to treatment onset exhibited similar fertility status to the normal cows (Jeremejeva *et al.* 2012).

h. Iodine *Versus* Antibiotics

In a recent study (Cengiz *et al.* 2022) i.u iodine administration was found to be superior to antibiotics for reconception after late-embryonic losses in dairy cows. The study compared conception and insemination efficiency responses to i.u infusion of 50 mL of normal saline, polyvinylpyrrolidone-iodine (50 mL; 2% v/v) or 150 mg of amoxicillin trihydrate and 40 mg/mL gentamicin sulphate in cows that were 196±28 days in milk. All cows followed the progesterone-based fixed-time insemination protocol. The polyvinylpyrrolidone-iodine treatment (59.5%) was as effective as the antibiotics (59.4%), compared to the saline treatment (21.7%) to achieve conception. The cows in polyvinylpyrrolidone-iodine group conceived in a significantly shorter time than those in group give antibiotics (46.0±8.7 days *versus* 105.0±10.0 days) with a significantly more efficient insemination to conception ratio (2.32±0.43 *versus* 4.10±0.32). It was concluded that i.u polyvinylpyrrolidone-iodine administration is superior to i.u antibiotics administration in rescheduling reproductive protocol upon late embryonic and fetal losses. I.u iodine treatment could be more cost-effective than i.u antibiotics by eliminating the necessity of milk discharge and avoiding antibiotic resistance (Carleton *et al.* 2008). Mechanisms by which antiseptic and antibiotic affect uterine tissue are different. Although antibiotics directly act on possible bacterial growth in the endometrium, polyvinylpyrrolidone-iodine can lead to a new inflammation, the degradation of the affected endometrial epithelium, and regeneration (Nakao *et al.* 1988). Galvão *et al.* (2009b) also explained that endometrial inflammations were not always concurrent with a bacterial infection and antibiotics could be ineffective on subclinical inflammations. Irritant effects of antiseptics may augment uterine tone, blood flow, mucus, and defense mechanisms (Bretzlaff, 1987), which are important to prepare the uterus for reconception in cows with late embryonic and early fetal death. Yoshida *et al.* (2020) reported significantly improved fertility after i.u polyvinylpyrrolidone-iodine infusion, which caused transient uterine inflammation that lasted for 2 d. Moreover, the same researchers reported earlier (48 h following administration) endometrial cell regeneration in the polyvinylpyrrolidone-iodine treatment compared to saline solution.

i. Formosulphathiazole *Versus* Cephapirin

Formosulphathiazole is a long-chain sulfonamide, derived from the condensation of sulphatiazole and formaldehyde, effective against a wide range of bacteria and viruses (Brander *et al.* 1992). This molecule is used successfully to treat bacterial enterocolitis in human and animals (Bagchi, 1951; Lammler and Durr, 1967) because of it's low intestinal absorption and good clinical effectiveness. Because the competition of sulfonamides with PABA is quantitative, the presence of excessive amounts of the latter, as in the presence of tissue exudates, pus or necrotic material, may result in reduction or cancelation of sulfa activity (Carli and Villa, 2008). In the postpartum uterus, with necrotic tissue debris and dead leukocytes, most of the metabolites derived through para amino benzoic acid pathways are readily available for the bacteria; consequently, sulfonamides could be ineffective (Olson *et al.* 1986). Because of the mechanism of action and the importance of protein synthesis during bacterial growth, the efficiency of sulfonamides is higher during the acute phase of disease and after metabolite stocks have been exhausted (Carli and Villa, 2008). As in the intestine, formosulphathiazole is poorly absorbed in the uterine lumen, persisting at high concentrations for a long period, and could, therefore, be very efficient in purulent infections (Dobson and Noakes, 1990). Earlier, the latter authors used formosulphathiazole to prevent uterine infections in cows after parturition; the authors showed that in treated cows there was a significant reduction of uterine infections caused by *A. pyogenes* and pathologic vaginal discharges. However, there is no information on its efficacy in treating clinical endometritis.

Mari *et al.* (2012) compared the clinical and bacteriologic recovery following a single i.u administration of 2500 mg of formosulphatiazole, 500 mg of cephapirin or placebo (4250 mg of propylene glycol) in cows ($\geq$ 28 days postpartum) with clinical endometritis. Cows were examined at the first estrus after treatment or no more than 30 days after treatment. Bacteria isolated were *E. coli*, *A. pyogenes*, *Pasteurella* spp. and *Streptococcus* spp. After treatment, 20.0%, 25.0% and 73.1% showed a positive bacteriologic culture. The clinical endometritis score after treatment also followed a similar pattern. It was concluded that the commercial formosulphatiazole preparation was as effective as cephapirin and more effective than placebo for the treatment of clinical endometritis in cows.

CHAPTER 4

EPILOGUE

Pankaj Sood

Uterine infections have a prevalent occurrence among cows, resulting in diminished welfare and fertility, as well as a significant economic impact. Because the causes of uterine diseases are complex and multifactorial, a holistic approach merits consideration when trying to identify or prevent them. The state of the uterus, immune system, lactation, and body condition of fresh postpartum cows is vastly different from that of cows that are regularly cycling and extend beyond the postpartum period. Furthermore, there are several other peripartum situations, such as obstetrical or metabolic issues, that make the fresh postpartum cows more vulnerable to setting up uterine infections. Therefore, the diagnostic and treatment approaches, as well as the treatment outcomes, may differ for the aforementioned two types of cows. Moreover, an increased prevalence of uterine infections in cows compared to small ruminants and other domesticated animals suggests the role of inappropriate management practices. The detection of abnormal uterine or genital discharge is regarded as the most convenient indicator of uterine infections. However, depending on the stage of the estrus and the involvement of other segments of the reproductive tract, such as the vagina and the cervix, which may not necessarily involve the uterus, the reliability of abnormal genital discharge as the sole indicator of uterine infections is reduced. The subclinical manifestation of uterine disease, whose incidence may exceed that of the clinical forms, is arduous to reliably detect using conventional methods, particularly the genital discharge, which appears to be normal. Subclinical forms of uterine diseases are equally detrimental to clinical forms in impairment of reproductive processes. Almost all dairy cows are infected with microbes after calving, which is unique to the dairy cow. Diverse arrays of microbial pathogens have been identified in the uterus of infected cows. With advancements in diagnostic modalities, the role of uterine microbes detected using conventional microbiological techniques has been a point of debate for quite some time. For example, it has been questioned whether opportunistic pathogens can cause uterine disease. Regardless of the

DOI: 10.4324/9781003749257-4

underlying cause, uterine infection damages the uterus, alters the uterine microenvironment, and may even involve the hypothalamo-pituitary-ovarian axis. These factors have a significant impact on the gametes and have the potential to affect fertilization, implantation, and embryonic and fetal development. The administration of antibiotics, whether intrauterine, systemically, or a combination thereof, has been in vogue in the treatment of uterine infections. However, due to several limitations of antibiotics, including legal considerations and public health significance in lactating animals, numerous other non-antibiotic therapeutic strategies have been investigated and yielded varying success rates. Since, most treatments have a definite cost linked to them and have an unpredicted outcome, efforts should be focused on management strategies to reduce peripartal problems and prevent the risk factors that predispose cows to uterine infections.

REFERENCES

Aarestrup F. 2012. Sustainable farming: Get pigs off antibiotics. *Nature* 486:465-466.

Abd El-Kader HA and Shehata SH. 2001. Bacteriological evaluation of vaginal discharges in cows with endometritis and clinically healthy heifers in Assiut governorate. *AUBER* 4 (2).

Abdel MA, Emad S, Gohary EL, Gabr AA, Ahmed SA, Ebrahim SA and Mohamed MF. 2016. Impact of supplementing propylene glycol and calcium propionate to primiparous buffalo cows during the late gestation and early lactation period on reproductive performance and metabolic parameters. *Alex J Vet Sci* 51:114-121.

Adnane M and Chapwanya A. 2022. A Review of the diversity of the genital tract microbiome and implications for fertility of cattle. *Animals* 12:460.

AHFS Drug Information. 2010. Acetylcysteine, Ed; McEvoy GK, 42nd edition, American Society of Health-System Pharmacists, USA, pp: 3652-3655.

Ahmadi MR, Haghkhah M, Derakhshandeh A, Aghamiri SM, Mirzaei A, Nazifi S and Makki M. 2015. Identification of bacterial and fungal agents of clinical endometritis in dairy heifers and treatment by metronidazole or cephapirin. *Insight* 5:99-112.

Ahmadi MR, Makki M and Mirzaei A. 2019. Effects of hypertonic dextrose and paraffin solution as nonantibiotic treatments of clinical endometritis on reproductive performance of high producing dairy cows. *Reprod Dom Anim* 54:762-771.

Ahmadi, MR, Khodakaram Tafti A, Nazifi S and Ghaisari HR. 2005. The comparative evaluation of uterine and cervical mucosa cytology with endometrial histopathology in cows. *Comp Clin Path.* 14:90-94.

Aknazarov BK. 1988. Study of humoral factors of local protective function in cows. *Veterinariya* 8:4143.

Alajbegovic S, Sanders JW, Atherly DE and Riddle MS. 2012. Effectiveness of rifaximin and fluoroquinolones in preventing travelers' diarrhea (TD): a systematic review and meta-analysis. *Syst Rev* Aug 28:39.

Alam M, Sinha AK and Banerjee NC. 1990. Biokinetic appraisal of ampicillin in cows. *Indian J Anim Sci* 60:562-564.

Alam M. 1988. Pharmacokinetics of ampicillin in buffalo calves, goat and poultry. Ph.D. Dissertation, Birsa Agricultural University, Ranchi.

Albihn A, Gustafsson H, Hurst M and Rodriguez-Martinez H. 1991. Embryonic ability to prolong the interoestrous interval in virgin and repeat breeder heifers. *Anim Reprod Sci* 26:193-210.

Al-Eknah MM and Noakes DE. 1989. Uterine activity in cows during the oestrous cycle, after ovariectomy and following exogenous oestradiol and progesterone. *Br Vet J* 145:328-336.

Alkholi UM, Al-Monem NA, El-Azim AAA and Sultan MH. 2011. Serum procalcitonin in viral and bacterial meningitis. *J Glob Infect Dis* 3:14-18.

Allen WE. 1978. Plasma concentrations of sodium benzylpenicillin after intrauterine infusion in pony mares. *Equine Vet* J 10:171-173.

Almeida AP, Ayalon N and Bartoov B. 1986. Bovine endometrial ultrastructure 6 and 7 days post-breeding. *Anim Reprod Sci* 10:293-300.

Alsemgeest SPM, Kalsbeek HC, Wensing T, Koeman JP, Van Ederen AM and Gruys E. 1994. Concentrations of SAA (SAA) and haptoglobin (Hp) as parameters of inflammatory diseases in cattle. *Vet Q* 16:21-23.

Alsemgeest SPM, Vant Clooster GAE, Van Miert ASJ, Hulskamp-Koch CK and Gruys E. 1996. Primary bovine hepatocytes in the study of cytokine induced acute-phase protein secretion *in vitro*. *Vet Immunol Immunopathol* 53:179-184.

Alsemgeest, SPM, Horadagoda A, Hulskamp-Koch CK, Tooten PCJ, Kim D H, Niewold THA and Gruys E 1995. First evidence for the existence of multiple isoforms of bovine serum amyloid-A. *Scand J Immunol* 41:407-413.

Amann RI, Ludwig W and Schleifer KH. 1995. Phylogenetic identification and in situ detection of individual microbial cells without cultivation. *Microbiol Rev* 59:143-69.

Amri A, Yaakub H and Johari A. 2015. The effects of different timing of artificial insemination, the association of uterine tone and site of semen deposition on pregnancy rate following oestrussynchronisation in beef cattle. *JTAFS* 43:137-144.

Anderson KL, Hamida NA, Frank A, Whitmore HL and Gustafsson BK. 1985. Collection and phagocytic evaluation of uterine neutrophilic leucocytes. *Theriogenology* 24:205–315.

Anika TT, Al Noman Z, Ferdous MRA, Khan SH, Mukta MA, Islam MS, Hossain MT, Rafiq K. 2019. Time dependent screening of antibiotic residues in milk of antibiotics treated cows. *J Adv Vet Anim Res*. 26:516-520.

Anonymous 2019. European Medicine Agency. Full AMEG Report: Categorisation of Antibiotics in the European Union Answer to the Request from the European Commission for Updating the Scientific Advice on the Impact on Public Health and Animal Health of the Use of Antibiotics in Animals, EMA/CVMP/CHMP/682198/2017. 2019. Available online: https://www.ema.europa.eu/en/documents/report/categorisation-antibiotics-european-union-answer-request-europeancommission-updating-scientific_en.pdf (accessed on 14 February 2022).

Ansari-Lari M, Nazifi S, Rezaei M and Asadi-Fardaqi J. 2008. Comparative study of plasma proteins including haptoglobin and serum amyloid A in different types of traumatic reticuloperitonitis in cattle. *Comp Clin Pathol* 17:245.

Appiah MO, Wang J and Lu W. 2020. Microflora in the reproductive tract of cattle: a review. *Agriculture* 10:232.

Arango-Sabogal JC, Dubuc J, Krug C, Denis-Robichaud J and Dufour S. 2019. Accuracy of leukocyte esterase test, endometrial cytology and vaginal discharge score for diagnosing postpartum reproductive tract health status in dairy cows at the moment of sampling, using a latent class model fit within a Bayesian framework. *Prev Vet Med* 162:1-10.

Arbeiter K, Awad-Maselmeh M, Kopschitz MM *et al.* 1976. Uterusgenebespiegel- und Blutserumpiegel-bestimmungen bei der Stute nach parenteral er Verabreichung von Penicillin und Ampicillin. *Wien tierarztl Mschr* 63:298-304.

Archer H, Barnett S, Irving S, Middleton K and Seal D. 1990. A controlled model of moist wound healing: Comparison between semi permeable film, antiseptics and sugar paste. Journal of Experimental Pathology (Oxford, England), 71:155.

Arias LAQ, Fernandez MV, Gonzalez JJB, Lopez MB, Herradon PJG and Martinez AIP. 2018. Subclinical endoemetritis in dairy cattle. Chapter 5. In: *New Insights into Theriogenology*. Intech Open. http://dx.doi.org/10.5772/intechopen.80229.

Aronowski J and Zhao X. 2011. Molecular pathophysiology of cerebral hemorrhage: secondary brain injury. *Stroke* 42:1781-1786.

Arrioja A. 2001. Product monographs, in Arrioja A (ed). Compendium of Veterinary Products, ed 6. Port Huron, MI, Adrian J. Bayley. Roncada, P.; Ermini, L.; Schleuning, A.; Stracciari, G.L.; Strocchia, A. Pharmacokinetics and residual behaviour in milk of oxytetracycline in cows following administration of uterine pessaries. *J Vet Pharmacol Ther* 2000, 23:281–285.

Arthur GH, Noakes DE and Pearson H. 2009. In: *Veterinary Reproduction and Obstetrics*. 9th edn., Bailliere Tindall, London, pp. 384-389.

Asbury AC and Hansen PJ. 1987. Effects of susceptibility of mares to endometritis and stage of cycle on phagocytic activity of uterine-derived neutrophils. *J Reprod Fertil Suppl* 35:311–316.

Asbury AC, Hallwell IW and Foster GW. 1980. Immunoglobulins in uterine secretions of mares with differing resistance to endometritis. *Theriogenology* 14:299–308.

Asbury AC. 1984. Uterine defense mechanisms in the mare: the use of intrauterine plasma in the management of endometritis. *Theriogenology* 21:387-393.

Asselin E, Goff AK, Bergeron H and Fortier MA. 1996. Influence of sex steroids on the production of prostaglandins F2α and E2 and response to oxytocin in cultured epithelial and stromal cells of the bovine endometrium. *Biol Reprod* 54:371-379.

Assey RJ, Purwantara B, Greve T, Hyttel P and Schmidt MH. 1993. Corpus luteum size and plasma progesterone levels in cattle after cloprostenol-induced luteolysis. *Theriogenology* 39:1321-1330.

Atkins JA, Smith MF, MacNeil MD, Jinks EM, Abreu FM, Alexander LJ and Geary TW. 2013. Pregnancy establishment and maintenance in cattle. *J Anim Sci* 91:722-733.

Ault TB, Clemmons BA, Reese ST, Dantas FG, Franco GA, Smith TP, Edwards JL, Myer PR and Pohler KG. 2019a. Bacterial taxonomic composition of the postpartum cow uterus and vagina prior to artificial insemination. *J Anim Sci* 97:4305-4313.

Ault TB, Clemmons BA, Reese ST, Dantas FG, Franco GA, Smith TP, Edwards JL, Myer PR and Pohler KG. 2019b. Uterine and vaginal bacterial community diversity prior to artificial insemination between pregnant and nonpregnant postpartum cows. *J Anim Sci* 97:4298-4304.

Ayliffe TR and Noakes DE. 1978a. Intrauterine absorption of sodium benzyl pencillin in the cow. *J Vet Pharmacol Therap* 1:267-271.

Ayliffe TR and Noakes DE. 1978b. Some preliminary studies on the uptake of sodium benzyl penicillin by the endometrium of cows. *Vet Rec* 102:215-217.

Ayliffe TR and Noakes DE. 1982. Effects of exogenous oestrogen and experimentally induced endometritis on absorption of sodium benzylpenicillin from the cow's uterus. *Vet Rec* 110:96-98.

Azawi OI. 2008. Postpartum uterine infection in cattle. *Anim Reprod Sci* 105:187-208.

Badinand F. 1975. Les metritis chez lavacche, influence des facteurs hormonaux et nutritionnels. *Cah Med Vet* 44:205-221.

Bagchi AK. 1951. Formo-sulphathiazole in acute enterocolitis—a report of 30 cases. *J Indian Med Assoc* 20:446-450.

Bage R, Gustafsson H, Larsson B, Forsberg M and Rodrýguez-Martýnez H. 2002. Repeat breeding in dairy heifers: follicular dynamics and estrous cycle characteristics in relation to sexual hormone patterns. *Theriogenology* 57:2257-2269.

Baggot JD, Love DN, Rose RJ and Raus R. 1985. Selection of an aminoglycoside antibiotic for administration to horses. *Equine Vet J* 19:195-197.

Baishaya SK, Das KK, Borgohain BN and Rahman H. 1999. Clinical trials with some antimicrobial agents in repeat breeding crossbred cattle. *Indian J Anim Reprod* 20:167-168.

Ball B, Shin S, Patten V, Lein D, Woods G. 1988. Use of a low-volume uterine flush for microbiologic and cytologic examination of the mare's endometrium. *Theriogenology* 29:1269-1283.

Ball L, Olson JD and Mortimer RG. 1984. Therapeutic considerations for postpartum bovine uterus. *Newsletter Soc Theriog* 7(1).

Ballas P, Reinländer U, Schlegl R, Ehling-Schulz M, Drillich M and Wagener K. 2021. Characterization of intrauterine cultivable aerobic microbiota at the time of insemination in dairy cows with and without mild endometritis. *Theriogenology* 159:28-34.

Banerjee S, Jia Y, Parker Siburt CJ, Abraham B, Wood F, Bonaventura C, Henkens R, Crumbliss AL and Alayash AI. 2012. Haptoglobin alters oxygenation and oxidation of hemoglobin and decreases propagation of peroxide-induced oxidative reactions. *Free Radic Biol Med* 53:1317-1326.

Barañski W, Podhalicz-Dziêgielewska M, Zduñczyk S and Janowski T. 2012. The diagnosis and prevalence of subclinical endometritis in cows evaluated by different cytologic thresholds. *Theriogenology* 78:1939-1947.

Barlund CS, Carruthers TD, Waldner CL and Palmer CW. 2008. A comparison of diagnostic techniques for postpartum endometritis in dairy cattle. *Theriogenology* 69:714-723.

Barrette Jr WC, Hannum DM, Wheeler WD and Hurst JK. 2002. General mechanism for the bacterial toxicity of hypochlorous acid: abolition of ATP production. *Biochemistry* 28:9172-9178.

Barrio M, Becerra JJ, Herradon PG, Sebastian F, Fernandez M and Quintela LA. 2014. Fertility in cows with endometritis diagnosed by cytology at day 30 postpartum. Proc 12th Cong AERA. Abstract.

Barsan N, Koziej D and Weimar U. 2007. Metal oxide-based gas sensor research: how to? *Sensor Actuat B-Chem* 121:18–35.

Basbug O, Yurdakul I and Yuksel M. 2020. Evaluation of aerum amyloid A and procalcitonin in some inflammatory diseases of cattle. *Kafkas Univ Vet Fak Derg* 26:397-402.

Bauknecht T, Krahe B, Rechenbach U, Zahradnik HP and Breckwoldt M. 1981. Distribution of prostaglandin E2 and prostaglandin F2α receptors in human myometrium. *Acta Endocrinol* 98:446-450.

Baumann H and Gauldie J. 1994. The acute phase response. *Immunol Today* 15:74-80.

Beagley JC, Whitman KJ, Baptiste KE and Scherzer J. 2010. Physiology and treatment of retained fetal membranes in cattle. *J Vet Intern Med* 24:261-268.

Beam SW and Butler WR. 1998. Energy balance, metabolic hormones, and early postpartum follicular development in dairy cows fed prilled lipid. *J Dairy Sci* 81:121-131.

Bearden HJ and Fuquay JW. 1997. Applied Animal Reproduction, fourth ed. Simon and Schuster Co., Upper Saddle River, NY, pp. 309-328.

Bedewy RB and Rahaway MA. 2019. Comparative study for detection of subclinical endometritis in local cows. *Adv Anim Vet Sci* 7:289-294.

Beggard AFJM, Vanden-Hafen MJ and Kriele CPMA. 1992. Rationale for treatment of retained placenta in cows with neomycin metronidazole. *Vet Rec* 130:349-350.

Behera S, Chandrashekaramurthy V, Krishnaswamy A, Kumar G, Rao S and Nagaraj BN. 2015. Studies on bacterial profile of repeat breeding cows with subclinical endometritis. *Int J Sci Res* 6:2090-2092.

Bengtsson B and Greko C. 2014. Antibiotic resistance—Consequences for animal health, welfare, and food production. *Upsala J Med Sci* 119:96-102.

Benzaquen ME, Risco CA, Archbald LF, Melendez P, Thatcher MJ, and Thatcher WW. 2007. Rectal temperature, calving-related factors, and the incidence of puerperal metritis in postpartum dairy cows. *J Dairy Sci* 90:2804-2814.

Bernardi S, Rinaudo A and Marini P. 2016. Cervical mucus characteristics and hormonal status at insemination of Holstein cows. *Iranian J Vet Res* 17: p 45.

Bertics SJ, Grummer RR, Cadorniga-Valino C and Stoddard EE. 1992. Effect of prepartum dry matter intake on liver triglyceride concentration and early lactation. *J Dairy Sci* 75:1914-1922.

Bertoni G, Trevisi E, Han X and Bionaz M. 2008. Effects of inflammatory conditions on liver activity in puerperium period and consequences for performance in dairy cows. *J Dairy Sci* 91:3300-3310.

Beyene T. 2016. Veterinary drug residues in food-animal products: its risk factors and potential effects on public health. *J Vet Sci Technol* 7:1-7.

Bhat FA, Bhattacharyya HK and Hussain SA. 2014. White side test: A simple and rapid test for evaluation of nonspecific bacterial genital infections of repeat breeding cattle. *Vet Res Forum* 5:177-180.

Bhatt MK, Shankar MB, Saluja AK, Dholwani KK and Captain AD. 2012. Evaluation of antibacterial activ-ity of Ocimum sanctum methanolic extract. *J Pharm Sci Innov* 1:39-41.

Bhatt VK, Inderjeet, Verma SK and Gupta RC. 1988a. Disposition of ampicillin in healthy and diseased diestrous buffalo uteri upon local and systemic administration. *11th International Congress on Animal Reproduction and AI*. Dublin, Ireland. 1:502.

Bhatt VK, Inderjeet, Verma SK and Gupta RC. 1988b. Disposition of gentamicin in normal and diseased buffalo uteri following intramuscular and intrauterine administration. *11th International Congress on Animal Reproduction and AI*, Dublin, Ireland. 1:503.

Bicalho ML, Machado VS, Higgins CH, Lima FS and Bicalho RC. 2017. Genetic and functional analysis of the bovine uterine microbiota. Part I: Metritis versus healthy cows. *J Dairy Sci* 100:3850-3862.

Bicalho ML, Machado VS, Oikonomou G, Gilbert RO and Bicalho RC. 2012. Association between virulence factors of *Escherichia coli*, *Fusobacterium necrophorum*, and *Arcanobacterium pyogenes* and uterine diseases of dairy cows. *Vet Microbiol* 157:125-131.

Bicalho MLS, Lima FS, Machado VS, Meira Jr EB, Ganda EK, Foditsch C, Bicalho RC and Gilbert RO. 2016. Associations among *Trueperella pyogenes*, endometritis diagnosis, and pregnancy outcomes in dairy cows. *Theriogenology* 85:267-274.

Bicalho RC, Machado VS, Bicalho ML, Gilbert RO, Teixeira AG, Caixeta LX and Pereira RV. 2010. Molecular and epidemiological characterization of bovine intrauterine *Escherichia coli*. *J Dairy Sci* 93:5818-5830.

Bieniek K, Szuster-Ciesielska A, Kaminska T, Kondracki M, Witek M, Kandefer-Szerszen M, 1998: Tumor necrosis factor and interferon activity in the circulation of calves after repeated injection of low doses of lipopolysaccharide. *Vet Immunol Immunopathol* 62:297-307.

Bierschwal CJ and Uren AW. 1958. The absorption of chlortetracycline (Aureomycin) by the bovine uterus. *J Am Vet Med Assoc* 129:373-374.

Bierschwal CJ, Dale HE and Uren AW. 1955. The absorption of sulfamethazine by the bovine involuting uterus. *J Am Vet Med Assoc* 126:398-399.

Bilk S, Nordhoff M, Schulze C, Wieler LH and Kutzer P. 2011. Antimicrobial susceptibilities and occurrence of resistance genes in bovine Helcococcus ovis isolates. *Vet Microbiol* 149:488-491.

Black WD, Holt JD and Gentry RD. 1983. Pharmcokinetic study of neomycin in calves following intravenous and intramuscular administration. *J Comp Med* 47:433-435.

Blair GS, Folley SJ, Malpress FH and Coppen FMV. 1941. Variations in certain properties of bovine cervical mucus during the oestrous cycle. *Biochem J* 35: p 1039.

Bloch A, Folman Y, Kaim M, Roth Z, Braw-Tal R and Wolfenson D. 2006. Endocrine alterations associated with extended time interval between estrus and ovulation in high-yield dairy cows. *J Dairy Sci* 89:4694-4702.

Blue MG, Brady AA, Davidson JW and Kenney RK. 1982. Studies on the composition and antibacterial activity of uterine fluid in mare. *J Reprod Fertil* (Suppl) 32:143-146.

Blum JW, Dosogne H, Hoeben D, Vangroenweghe F, Hammon HM, Bruckmaier RM and Burvenich C. 2000. Tumor necrosis factor-α and nitrite/nitrate responses during acute mastitis induced by Escherichia coli infection and endotoxin in dairy cows. *Dom Anim Endocrinol* 19:223-235.

Bohn AA, Ferris RA and McCue PM. 2014. Comparison of equine endometrial cytology samples collected with uterine swab, uterine brush, and low volume lavage from healthy mares. *Vet Clin Pathol* 43:594-600.

Boitor I, Muntean M, Mates N, Moise D, Morar R, Boitor M, *et al.*, 1980, Laboratory and therapeutic studies of puerperal endometritis and repeat breeding in the cow. *Bulletin Institutului Agronomic Cluj Napoca, Zootehnie si Medicina Veterinary* 34:111-114.

Bolinder A, Seguin B, Kindahl H, Bouley D and Otterby D. 1988. Retained fetal membranes in cows: manual removal versus non-removal and its effect on reproductive performance. *Theriogenology* 30:45-56.

Bollo E, Biolatti B, Pau S and Galloni M. 1990. Scanning electron microscopy of pathogenic changes in the epithelial surfaces of the uterus and uterine tubes of cows. *Am J Vet Res* 51:137-142.

Bonafos LD, Kot K and Ginther OJ. 1995. Physical characteristics of the uterus during the bovine estrous cycle and early pregnancy. *Theriogenology* 43:713-721.

BonDurant RH. 1999. Inflammation in the bovine reproductive tract. *J Dairy Sci* 82(Suppl. 2):101-110.

Bonnett BN, EtheringtonWG, Martin SW and JohnsonWH. 1990. The effect of prostaglandin administration to Holstein Friesian cows at day 26 postpartum on clinical findings, and histological and bacteriological results of endometritis biopsies at day 40. *Theriogenology* 33:877-890.

Bonnett BN, Martin SW and Meek AH. 1993. Associations of clinical findings, bacteriological and histological results of endometrial biopsy with reproductive performance of postpartum dairy cows. *Prev Vet Med* 15:205-220.

Bonnett BN, Martin SW, Gannon VP, Miller RB and Etherington WG. 1991a. Endometrial biopsy in Holstein-Friesian dairy cows. III. Bacteriological analysis and correlations with histological findings. *Canadian J Vet Res* 55: p 168.

Bonnett BN, Miller RB, Etherington WG, Martin SW and Johnson WH. 1991b. Endometrial biopsy in Holstein-Friesian dairy cows. I. Technique, histological criteria and results. *Can J Vet Res* 55:155-161.

Bonnett BN, Miller RB, Martin SW, Etherington WG and Buckrell BC. 1991c. Endometrial biopsy in Holstein-Friesian dairy cows. II. Correlations between histological criteria. *Can J Vet Res* 55:162-167.

Boothe DM. 2001. Drugs affecting the respiratory system. In: *The Veterinary Pharmacology and Therapeutics*, Ed; Adams HR, 8th edition, Iowa State Press, Ames, Iowa, USA, pp; 1105-1119.

Borowieck J. 1989. Therapeutic trial with an alcoholic plant extract in bovine puerperal endometritis. *Medyena Weteryaryjna* 45:221-224.

Borsberry S and Dobson H. 1989. Periparturient diseases and their effect on reproductive performance in five dairy herds. *Vet Rec* 124:217-219.

Bosch A, Prieto C, Serra DO, Martina P, Stammbler M, Naumann D, Schmitt J and Yantorno O. 2010. Type-IV pili spectroscopic markers: applications in the quantification of piliation levels in Moraxella bovis cells by a FT-IR ANN-based model. *J Biophoton* 3:522-533.

Bosu WT, Liptrap RM and Leslie KE. 1984. Peripartal changes in plasma progesterone and 15-keto-13,14-dihydro-prostaglandin F2 concentrations in Holstein cows with and without retained fetal membranes. *Anim Reprod Sci* 7:497-510.

Bourke M, Mills JN and Barnes A. 1997. Collection of endometrial cells in the mare. *Aust Vet J* 75:755-758.

Bouters R and Vandeplasche M. 1977. Postpartum infection in cattle: diagnosis and prevention. *J S Afr Vet Assoc* 48:237-239.

Boyd EH and Allen WE. 1988a. Absorption of neomycin from the equine uterus: effect of bacterial and chemical endometritis. *Vet Rec* 122:37-39.

Boyd EH and Allen WE. 1988b. Absorption of neomycin from the postpartum equine uterus. *Equine Vet J* 20:63-65.

Brander GC, Pugh DM, Bywater RJ and Jenkins WL. 1992. Veterinary applied pharmacology and therapeutics, 5th ed. Tindall: Baillière.

Brandt GW and Manning JP. 1996. Improved uterine biopsy technics for diagnosing infertility in the mare. *VM, SAC*. 4:977-983.

Bréchard S and Tschirhart EJ. 2008. Regulation of superoxide production in neutrophils: Role of calcium influx. *J Leukoc Biol* 84:1223-1237.

Breijyeh Z, Jubeh B and Karaman R. 2020. Resistance of gram-negative bacteria to current antibacterial agents and approaches to resolve it. *Molecules* 25:1340.

Bretzlaff K. 1987. Rationale for treatment of endometritis in dairy cows. *Vet Clin North Am Food Anim Prac* 3:593–598.

Bretzlaff KN, Ott RS, Koritz GD *et al.* 1982. Distribution of oxytetracycline in the genital tract of cows. *Am J Vet Res* 43:12-16.

Bretzlaff KN, Ott RS, Koritz GD *et al.* 1983a. Distribution of oxytetracycline in genital tract tissues of postpartum cows given the drug by intravenous and intrauterine routes. *Am J Vet Res* 44:764-769.

Bretzlaff KN, Ott RS, Koritz GD *et al.* 1983b. Distribution of oxytetracycline in the healthy and diseased postpartum genital tract of cows. *Am J Vet Res* 44:760-763.

Bretzlaff KN, Ott RS, Koritz GD, Lock TF, Bevill RF, Shawley RV, Gustafsson BK and Davis LE. 1982. Distribution of oxytetracycline in the genital tract of cows. *Am J Vet Res* 43:12-16.

Bretzlaff KN, Ott RS, Koritz GD, Lock TF, Neff-Davis CA, Gustafsson BK and Davis LE. 1988. Distribution of chloramphenicol in the genital tract of postpartum cows. *Am J Vet Res* 49:914-917.

Bretzlaff KN. 1986. Distribution of chloramphenicol in the bovine genital tract and pharmacokinetic studies on florfenicol in cattle. *Dissertation Abst Int B* 47:957.

Brick T, Schuenemann G, Bas S, Daniels J, Pinto C, Rings D and Rajala Schultz, P. 2012. Effect of intrauterine dextrose or antibiotic therapy on reproductive performance of lactating dairy cows diagnosed with clinical endometritis. *J Dairy Sci* 95:1894-1905.

Bridges GA, Mussard ML, Burke CR and Day ML. 2010. Influence of the length of proestrus on fertility and endocrine function in female cattle. *Anim Reprod Scie* 117:208-215.

Briel M, Schuetz P, Mueller B, Young J, Schild U, Nusbaumer C, Periat P, Bucher HC and Christ-Crain M. 2008. Procalcitonin-guided antibiotic use *vs* a standard approach for acute respiratory tract infections in primary care. *Arch Intern Med* 168:2000-2007.

Brodzki P, Kostro K, Krakowski L and Marczuk J. 2015. Inflammatory cytokine and acute phase protein concentrations in the peripheral blood and uterine washings of cows with subclinical endometritis in the late postpartum period. *Vet Res Commun* 39:143-149.

Brook D. 1993. Uterine cytology. Equine Reproduction Lea and Febiger. Philadelphia . p. 246-54.

Brown MP, Emberston RM, Gronwall RR, Beal C, Mayhew IG and Curry SH. 1984. Amikacin sulfate in mares: Pharmacokinetics and body fluid and endometrial concentrations after repeated intramuscular administration. *Am J Vet Res* 45:1610-1613.

Bruins M, Bos A, Petit PL, Eadie K, Rog A, Bos R, van Ramshorst GH and van Belkum A. 2009. Device-independent, real-time identification of bacterial pathogens with a metal oxide-based olfactory sensor. *Eur J Clin Microbiol Infect Dis* 28:775-780.

Bruins M, Rahim Z, Bos A, van de Sande WWJ, Endtz HP and van Belkum A. 2013. Diagnosis of active tuberculosis by e-nose analysis of exhaled air. *Tuberculosis* 93:232-238.

Brunkhorst FM, Heinz U and Forycki ZF. 1998. Kinetics of procalcitonin in iatrogenic sepsis. *Intensive Care Med* 24:888-889.

Brus DHJ. 1954. Biopsia Uteri, Thesis, University of Utrecht, Utrecht, Netherlands.

Bu J. 1955. Studies on local treatment of the uterus in the estral and luteal phases of the cycle. *Nord Vet Med* 7:917 (Cited by Bretzlaff, 1986).

Buehler PW, Abraham B, Vallelian F, Linnemayr C, Pereira CP, Cipollo J F, Jia Y, Mikolajczyk M, Boretti FS, Schoedon G, Alayash AI and Schaer DJ. 2009. Haptoglobin preserves the CD163 hemoglobin scavenger pathway by shielding hemoglobin from peroxidative modification. *Blood* 113:2578-2586.

Bulman DC and Lamming GE. 1978. Milk progesterone levels in relation to conception, repeat breeding and factors influencing acyclicity in dairy cows. *Reproduction* 54:447-458.

Burfeind O, Bruins M, Bos A, Sannmann I, Voigtsberger R and Heuwieser W. 2014a. Diagnosis of acute postpartum metritis by electronic nose device analysis of vaginal discharge. *Theriogenology* 82:64-70.

Burfeind O, Sannmann I, Voigtsberger R and Heuwieser W. 2014b. Receiver operating characteristic curve analysis to determine the diagnostic performance of serum haptoglobin concentration for the diagnosis of acute puerperal metritis in dairy cows. *Anim Reprod Sci* 149:145-151.

Burfeind O, Suthar VS and Heuwieser W. 2012. Effect of heat stress on body temperature in healthy early postpartum dairy cows. *Theriogenology* 78:2031-2038.

Burfeind O, von Keyserlingk MA, Weary DM, Veira DM and Heuwieser W. 2010. Short communication: Repeatability of measures of rectal temperature in dairy cows. *J Dairy Sci* 93:624-627.

Burkhardt O, Merker H J, Shakibaei M and Lode H. 2003. Electron microscopic findings in BAL of a fire eater after petroleum aspiration. *Chest J* 124:398-400.

Burrows GE, Barto PB and Martin B. 1987. Comparative pharmacokinetics of gentamicin, neomycin and oxytetracycline in new born calves. *J Vet Pharmacol Therap* 10:54-63.

Burton NR and Lean IJ. 1995. Investigation by meta-analysis of the effect of prostaglandin F2a administered postpartum on the reproductive performance of dairy cattle. *Vet Rec* 136:90-94.

Busch W. 2018. Intrauterine Therapie der chronischen Endometritis des Rindes mit Peressigsäure [Treatment of chronic endometritis in cattle by intrauterine application of peroxyethan acid]. *Tierarztl Prax Ausg G Grosstiere Nutztiere.* 46:5-11.

Butler WR. 1998. Review: effect of protein nutrition on ovarian and uterine physiology in dairy cattle. *J Dairy Sci* 81:2533-2539.

Butt BM, Bessre TE and Senger PLPR. 1993. Specific antibody to Haemophilus somnus in the bovine uterus following intramuscular immunization. *Infect Immunol* 61:2558-2562.

Butt BM, Senger PL and Widders PR. 1991. Neutrophil migration into the bovine uterine lumen following intrauterine inoculation with killed *Haemophilus somnus*. *J Reprod Fertil* 93:341-345.

Cai TQ, Weston PG, Lund LA, Brodie B, McKenna DJ and Wagner WC. 1994. Association between neutrophil functions and periparturient disorders in cows *Am J Vet Res* 55:934-943.

Cairoli F, Battocchio M, Veronesi MC, Brambilla D, Conserva F, Eberini I, Wait R and Gianazza E. 2006. Serum protein pattern during cow pregnancy: Acute-phase proteins increase in the Peripartum period. *Electrophoresis* 27:1617-1625.

Cammarato G, Sanguinetti M, Gallo A, Posteraro B, 2012. Review article: Biofilm formation by *Helicobacter pylori* as a target for eradication of resistant infection. *Aliment Pharmacol Ther* 36:222-230.

Canete Soler R, Lopez Bernal A and Turnbull AC. 1987. Conversion of prostaglandin E2 to prostaglandin F2α by human myometrium. *Horm Metab Res* 19:515-516.

Carleton CL and Threfall WR. 1984. Determination of changes in the rate of absorption and histologic changes of endometrium following intrauterine Lugol's infusion of the postpartum cow. In: Proceedings of the 10th International Congress on Animal Reproduction and AI. Urbana, Champaign, III, pp. 412-418.

Carleton CL, Threlfall WR and Schwarze RA. 2008. Iodine in milk and serum following intrauterine infusion of Lugol's solution. *Int J Appl Res Vet Med* 6:121-129.

Carli S and Villa R. 2008. Sulfamidici e derivati pirimidinici. In: Carli S, editor. *Farmacologia Veterinaria* Idelson Gnocchi; p. 709-715.

Carli S, Perretta S, Monti F, Sonzogni O and Faustini R. 1982. Serum levels, tissue distribution and residues of neomycin following intramuscular administration to chicks. *J Vet Pharmacol Therap* 5:203-207.

Carneiro LC, Cronin JG and Sheldon IM. 2016. Mechanisms linking bacterial infections of the bovine endometrium to disease and infertility. *Reprod Biol* 16:1-7.

Carson RL,Wolfe DF, Klesius PH, Kemppainen RJ and Scanlan CM. 1988. The effects of ovarian hormones and ACTH on uterine defense to Corynebacterium pyogenes in cows. *Theriogenology* 30:91-97.

Castelli GP, Pognani C, Meisner M, Stuani A, Bellomi D, Sgarbi L. 2004. Procalcitonin and C-reactive protein during systemic inflammatory response syndrome, sepsis and organ dysfunction. *Crit Care*, 8:R234.

Castrucci G, Frigeri F, Cilli V, Donelli G, Ferrari M, Chicchinri U and Bordoni E. 1986. A study of herpesvirus isolated from dairy cattle with a history of reproductive disorders. *Comp Immunol Microbiol Infect Dis* 9:13-21.

Caudle AB, Purswell BJ, Williams DJ, Brooks P and Rourke JE. 1983. Endometrial levels of amikacin in the mare after intrauterine infusion of amikacin sulfate. *Theriogenology* 19:433-439.

Celik H and Ayar A. 2002. Clarythromycin inhibits myometrial contractions in isolated human myometrium independent from stimulus. *Physiol Res* 51:239-245.

Celik H, Ayar A and Sapmaz E. 2001. Effects of erythromycin on stretch-induced contractile activity of isolated myometrium from pregnant women. *Acta Obstet Gynecol Scand* 80:697-701.

Celik H, Ayar A, Tug N and Baltaci AK. 2002. Erythromycin inhibits prostaglandin F2alpha-induced contractions of myometrium isolated from non-pregnant rats. *Br J Obstet Gynecol* 109:1036-1040.

Cengiz M, Ozbek S and Chacher MFA. 2022. Intrauterine iodine administration is superior to antibiotics for reconception after late-embryonic losses in dairy cows. *Can J Vet Res* 86:74-77.

Ceron JJ, Eckersall PD and Martynez-Subiela S. 2005. Acute phase proteins in dogs and cats: current knowledge and future perspectives. *Vet Clin Pathol* 34:85-99.

Cerri RLA, Rutigliano HM, Lima FS, Araujo DB and Santos JEP. 2009. Effect of source of supplemental selenium on uterine health and embryo quality in high-producing dairy cows. *Theriogenology* 71:1127-1137.

Chacin MFL, Hansen PJ and Drost M. 1990. Effects of stage of the estrous cycle and steroid treatment on uterine immunoglobulin content and polymorphonuclear leukocytes in cattle. *Theriogenology* 34:1169-1184.

Chan JP, Chang CC, Hsu WL, Liu WB and Chen TH. 2010. Association of increased serum acute-phase protein concentrations with reproductive performance in dairy cows with postpartum metritis. *Vet Clin Pathol* 39:72-78.

Chapwanya A, Meade KG, Doherty ML, Callanan JJ and O'Farrellya C. 2013. Endometrial epithelial cells are potent producers of tracheal antimicrobial peptide and serum amyloid A3 gene expression in response to E. coli stimulation. *Vet Immunol Immunopathol* 151:157-162.

Chapwanya A, Meade KG, Doherty ML, Callanan JJ, Mee JF and O'Farrelly C. 2009. Histopathological and molecular evaluation of Holstein-Friesian cows postpartum: toward an improved understanding of uterine innate immunity. *Theriogenology* 71:1396-1407.

Chapwanya A, Meade KG, Narciandi F, Stanley P, Mee JF, Doherty ML, Callanan JJ and Farrely CO. 2010. Endometrial biopsy: A valuable clinical and research tool on bovine reproduction. *Theriogenology* 73:988-994.

Charles M S. 1988. Bacterial virulence factors. pathology of bacterial diseases and bacterial infection and immunity. *Introduction to Veterinary Bacteriology*. 1st edn. lowa State University Press, Ames, Iowa.

Chaudhary RK. 1985. Pharmacokinetic studies of cephaloridine in *Bubalus bubalis*. M.V.Sc. Thesis, Punjab Agricultural University, Ludhiana.

Chaudhary SK, Gupta RC and Uppal RP. 1987a. Disposition of trimethoprim-sulfamethoxazole combination following intrauterine administration in experimentally metritis condition in buffalo. *Theriogenology* 28:961-967.

Chaudhary SK, Gupta RC and Uppal RP. 1987b. Disposition of trimethoprim-sulfamethoxazole combination following intrauterine administration in experimentally metritis condition in buffalo. *J Vet Med Assoc* 34:703-706.

Chen C, Song X, Wei W, Zhong H, Dai J, Lan Z, Li F, Yu X, Feng Q, Wang Z, Xie H, Chen X, Zeng C, Wen B, Zeng L, Du H, Tang H, Xu C, Xia Y, Xia H, Yang H, Wang J, Wang J, Madsen L, Brix S, Kristiansen K, Xu X, Li J, Wu R, Jia H. 2017. The microbiota continuum along the female reproductive tract and its relation to uterine-related diseases. *Nat Commun* 8:875.

Chengappa MM, Maddux RL, Greer SC, Pincus DH and Giest LL. 1984. Isolation and identification of yeasts and yeast-like organisms from clinical sources. *J Clin Microbiol* 19:427-428.

Cheong SH, Nydam DV, Galvao KN, Croiser BM and Gilbert RO. 2011. Cow-level and herd-level risk factors for subclinical endometritis in lactating Holstein cows. *J Dairy Sci* 94:762-770.

Cheong SH, Nydam DV, Galvão KN, Crosier BM, Ricci A, Caixeta LS, Sper RB, Fraga M and Gilbert RO. 2012. Use of reagent test strips for diagnosis of endometritis in dairy cows. *Theriogenology* 77:858-864.

Chi Z, Liu R, You H, Ma S, Cui H and Zhang Q. 2014. Probing the *in vitro* cytotoxicity of the veterinary drug oxytetracycline. *PLoS One* 14:9(7):e102334.

Ciernia LA, Perry GA, Smith MF, Rich JJ, Northrop EJ, Perkins SD, Green JA, Zezeski AL and Geary TW. 2021. Effect of estradiol preceding and progesterone subsequent to ovulation on proportion of postpartum beef cows pregnant. *Anim Reprod Sci* 227: p 106723.

Claesson HE, Odlander B and Jacobsson PJ. 1992. Leukotriene B4 in the immune system. *Int J Immunopharmac* 14:441-449.

Clemmons BA, Reese ST, Dantas FG, Franco GA, Smith TP, Adeyosoye OI, Pohler KG and Myer PR. 2017. Vaginal and uterine bacterial communities in postpartum lactating cows. *Front Microbiol* 8: p 1047.

Cocchia N, Paciello O, Auletta L, Uccello V, Silvestro L, Mallardo K, Paraggio G and Pasolini MP. 2012. Comparison of the cytobrush, cottonswab, and low-volume uterine flush techniques to evaluate endometrial cytology for diagnosing endometritis in chronically infertile mares. *Theriogenology* 77:89-98.

Coe BL and Allrich RD. 1989. Relationship between endogenous estradiol-17β and estrous behavior in heifers. *J Anim Sci* 67:1546-1551.

Cohen R O, Bernstein M and Ziv G. 1995. Isolation and antimicrobial susceptibility of Actinomyces pyogenes recovered from the uterus of dairy cows with retained fetal membranes and post parturient endometritis. *Theriogenology* 43:1389-1397.

Conzuelo F, Ruiz-Valdepeñas Montiel V, Campuzano S, Gamella M, Torrente-Rodríguez RM, Reviejo AJ and Pingarron JM. 2014. Rapid screening of mul-tiple antibiotic residues in milk using disposable amperometric magnetosensors. *Anal Chim Acta* 820:32-38.

Cook DL, Parfet JR, Smith CA, Moss GE, Youngquist RS and Garverick HA. 1991. Secretory patterns of LH and FSH during development and hypothalamic and hypophysial characteristics following development of steroid-induced ovarian follicular cysts in dairy cattle. *Reproduction* 91:19-28.

Cooke RG, Ahmad N and Nicholson T. 1998. Effect of progesterone and estradiol-17 on oxytocin-induced PGF2_ release and endometrial oxytocin receptor concentrations in ovariectomized goats. *Prostag Oth Lipid M* 55:109-120.

Cooray R, Waller KP and Venge P. 2007. Haptoglobin comprises about 10% of granule protein extracted from bovine granulocytes isolated from healthy cattle. *Vet Immunol Immunopath* 119:310-315.

Cooray, R. 1994. Use of bovine myeloperoxidase as an indicator of mastitis in dairy cattle. *Vet Microbiol* 42:317-326.

Corbeil LB, Hall CE, Lein D, Corbeil RR and Duncan JR. 1976. Immunoglobulin classes in genital secretions of mycoplasma-infected and normal heifers. *Infect Immunol* 13:1595-1600.

Corbeil LB, Schurig JR, Duncan JR, Corbeil RR and Winter AJ. 1974. Immunoglobulin classes and biological function of *Campylobacter* (Vibrio) *fetus* antibodies in serum and cervicovaginal mucus. *Infect Immunol* 10:422-429.

Couto GB, Vaillancourt DH and Lefebvre RC. 2013. Comparison of a leukocyte esterase test with endometrial cytology for diagnosis of subclinical endometritis in postpartum dairy cows. *Theriogenology* 79:103-107.

Couto MA and Hughes JP. 1985. Intrauterine inoculation of a bacteria-free filtrate of Streptococcus zooepietimicus in clinically normal and infected mares. *J Eq Vet Sci* 5:81-86.

Crawford RG, Leslie KE, Bagg R, Disck CP and Duffield TF. 2005. The impact of controlled release capsules of monensin on postcalving haptoglobin concentrations in dairy cattle. *Can J Vet Res* 69:208-214.

Crivei IC, Cozma AP, Bors SI, Ruginosu E, Ro'ca P and Drugociu D. 2016. Antibiotics sensitivity of isolated bacteria from dairy cows with clinical endometritis. LucrariStiintifice-Universitatea de Stiinte Agricole a Banatului Timisoara, *Vet Med (Praha)* 49:26-32.

Cullens FM. 2005. Effects of the timing of initiation of fat supplementation on productive and reproductive response of periparturient dairy cows during summer. University of Florida.

Cunha F, Jeon SJ, Daetz R, Vieira-Neto A, Laporta J, Jeong KC, Barbet AF, Risco CA and Galvão KN. 2018. Quantifying known and emerging uterine pathogens, and evaluating their association with metritis and fever in dairy cows. *Theriogenology* 114:25-33.

Curtain CC, Clark BL and Dufty JH. 1971. The origins of the immunoglobulins in the mucus secretions of cattle. *Clin Exp Immunol* 8:335-344.

Curtis CR, Erb HN, Sniffen CJ, Smith RD, and Kronfeld DS. 1985. Path analysis of dry period nutrition, postpartum metabolic and reproductive disorders, and mastitis in Holstein cows. *J Dairy Sci* 68:2347-2360.

CVMP (Committee for Medicinal Products for Veterinary Use). 2009. Revised reflection paper on the use of 3rd and 4th generation cephalosporins in food producing animals in the European Union: Development of resistance and impact on human and animal health. Accessed August, 2023. http://www.ema.europa.eu/docs/en_GB/document_library/Scientific_guideline/2009/10/WC500004307.pdf.

CVMP. 1996. Acetyl cysteine. Summary report. EMEA/MRL/96. European Agency for the Evaluation of Medicinal Products, London, UK.

Dahiya S, Kumari S, Rani P, Onteru SK and Singh D. 2018. Postpartum uterine infection and ovarian dysfunction. *Indian J Med Res* 148(Supplementary 1): p S64.

Dalhoff A, Ambrose PG and Mouton JW. 2009. A long journey from minimum inhibitory concentration testing to clinically predictive breakpoints: Deterministic and probabilistic approaches in deriving breakpoints. *Infection* 37:296-305.

Dandona P, Nix D, Wilson MF, Aljada A, Love J, Assicot M and Bohuon C. 1994. Procalcitonin increase after endotoxin injection in normal subjects. *J Clin Endocrinol Metab* 79:1605-1608.

D'Armiento J, Dalal SS and Chada K. 1997. Tissue, temporal and inducible expression pattern of haptoglobin in mice. *Gene* 195:19-27.

Das KL, Misra PR, Kar BC and Roy PK. 1996. Aerobic bacterial isolates and their antibiogram in repeat breeding cows. *Indian Vet J* 73:900-902.

Davies D, Meade KG, Herath S, Eckersall PD, Gonzalez D, White JO, Conlan RS, Farrelly CO and Sheldon IM. 2008. Toll-like receptor and antimicrobial peptide expression in the bovine endometrium *Reprod Biol Endocrinol* 6:53

de Boer M, Buddle BM, Heuer C, Hussein H, Zheng T, LeBlanc SJ and McDougall S. 2015a. Associations between intrauterine bacterial infection, reproductive tract inflammation, and reproductive performance in pasture-based dairy cows. *Theriogenology* 83:1514-1524.

de Boer M, Heuer C, Hussein H and McDougall S. 2015b. Minimum inhibitory concentrations of selected antimicrobials against *Escherichia coli* and *Trueperella pyogenes* of bovine uterine origin. *J Dairy Sci* 98:4427-4438.

de Boer MW, LeBlanc SJ, Dubuc J, Meier S, Heuwieser W, Arlt S, Gilbert RO and McDougall S. 2014. Invited review: Systematic review of diagnostic tests for reproductive-tract infection and inflammation in dairy cows. *J Dairy Sci* 97:3983-3999.

De Bois CHW and Manspeaker JE. 1986. Endometrial biopsy in the bovine. In: Morrow DA (Ed) *Current Therapy in Theriogenology*. W B Saunders Co. Philadelphia pp 424- 427.

De Bois CHW. 1961. Endometritis en vruehtbaaheit het rund. University of Utrecht, Ph.D. Thesis, pp 124-128.

De Kruif A, Gunnink JW and De Bois CH. 1982. Diagnosis and treatment of postpartum endometritis in cattle. *Vlaams Diergeneeskd Tijdschr* 107:717-725.

Debertolis L, Gaetano M, Barbara M, Sabine M, Eleonora I and Heinrich B. 2015. Effects of an induced endometritis on uterine blood flow in cows evaluated by transrectal Doppler sonography. *J Vet Sci* 17:189-197.

Delèvaux I, André M, Colombier M, Albuisson E, Meylheuc F, Bègue RJ, Piette JC and Aumaître O. 2003. Can procalcitonin measurement help in differentiating between bacterial infection and other kinds of inflammatory processes? *Ann Rheum Dis* 62:337-340.

Denis-Robichaud J and Dubuc J. 2015. Determination of optimal diagnostic criteria for purulent vaginal discharge and cytological endometritis in dairy cows. *J Dairy Sci* 98:6848-6855.

Deori S. 2002. Use of bacterial modulins in treatment of endometritis cows. M.V.Sc. Thesis. I.V.R.I.

Desai SP, Sharma VK, Prajapati BI, Chauhan PM, Sutaria TV and Nakhashi HC. 2017. Bacterial Spectra of Cervical Mucus and Conception Rate in Endometritic Repeat breeder Crossbred Cows Following *E. coli* LPS Therapy. *Proteus* 2: p 6510.

Devakumar C and Kumar R. 2008. Total synthesis of Azadirachtin: A chemical odyssey. *Curr Sci Ind* 95:1-3.

Dhaliwal GS, Murray RD and Woldehiwet Z. 2001. Some aspects of immunology of the bovine uterus related to treatments for endometritis. *Anim Reprod Sci* 67:135-152.

Dhaliwal GS, Murray RD, Dobson H, Montgomery J and Ellis WA. 1996b. Presence of antigen and antibodies in serum and genital discharges of cows from dairy herds naturally infected with *Leptospira interrogans serovar hardjo*. *Res Vet Sci* 60:163-167.

Dhaliwal GS, Murray RD, Dobson H, Montgomery J, EllisWA and Baker J. 1996a. Presence of antigen and antibodies in serum and genital discharges of heifers after intrauterine inoculation of *Leptospira interrogans serovar hardjo*. *Res Vet Sci* 60:157-162.

Dhaliwal GS, Murray RD, Rees EM, Howard CV and Beech DJ. 2002. Quantitative unbiased estimates of endometrial gland surface area and volume in cycling cows and heifers. *Res Vet Sci* 73:259-265.

Dhaliwal GS, Singh J and Singh N. 1998. Treatment of repeat breeding in dairy cows — a comprehensive approach. In: Proceedings of the 20th World Buiatrics Congress. Sydney, pp. 673-676.

Dickinson SE. 2016. Effect of pre-ovulatory follicle size on oocyte transcript abundance in beef cows (Doctoral dissertation, University of Missouri-Columbia).

Dieleman SJ, Bevers MM, Van Tol HTM and Willemse AH. 1986. Peripheral plasma concentrations of oestradiol, progesterone, cortisol, LH and prolactin during the oestrous cycle in the cow, with emphasis on the peri-oestrous period. *Anim Reprod Sci* 10:275-292.

Dini P, Farhoodi M, Hostens M, Van Eetvelde M, Pascottini OB, Fazeli MH and Opsomer G. 2015. Effect of uterine lavage on neutrophil counts in postpartum dairy cows. *Anim Reprod Sci* 158:25-30.

Djuricic D, Vince S, Ablondi M, Dobranic T and Samardzija M. 2012. Effect of preventive intrauterine ozone application on reproductive efficiency in Holstein cows. *Reprod Dom Anim* 47:87-91.

Dobson DP and Noakes DE. 1990. Use of a uterine pessary to prevent infection of the uterus of the cow after parturition. *Vet Rec* 127:128-131.

Dohmen MJ, Joop K, Sturk A, Bols PEJ and Lohuis JACM. 2000. Relationship between intra-uterine bacterial contamination, endotoxin levels and the development of endometritis in postpartum cows with dystocia or retained placenta. *Theriogenology* 54:1019-1032.

Dohmen MJ, Lohuis JACM, Huszeniczea G, Nagy P and Gacs M. 1995. The relationship between bacteriological and clinical findings in cows with sub acute/ chronic endometritis. *Theriogenology* 43:1379-1388.

Dong R, Wan B, Lin S, Wang M, Huang J, Wu Y, Wu Y, Zhang N and Zhu Y. 2019. Procalcitonin and liver disease: A literature review. *J Clin Transl Hepatol* 7:51-55.

Donofrio G, Franceschi V, Capocefalo A, Cavirani S and Sheldon IM. 2009. Isolation and Characterization of Bovine Herpesvirus 4 (BoHV-4) from a Cow Affected by Post-partum Metritis and Cloning of the Genome as a Bacterial Artificial Chromosome. *Reprod Biol Endocrinol* 19:83.

Dornbusch HJ, Strenger V, Kerbl R, Lackner H, Schwinger W, Sovinz P and Urban C. 2005. Procalcitonin-A marker of invasive fungal infection? *Support Care Cancer* 13:343-346.

Douglas SD: Cells involved in immune responses. 1980 In: Fudenberg HH, Stiies DP, Caklwell JL, Welfs JV, eds. *Basic and clinical immunology*, 3rd ed. Los Altos, California: Lange Medical Publications, 98:96-114.

Drillich M, Voigt D, Forderung D and Heuwieser W. 2007. Treatment of acute puerperal metritis with flunixin meglumine in addition to antibiotic treatment. *J Dairy Sci* 90:3758-3763.

Drillich M and Wagener K. 2018. Pathogenesis of uterine diseases in dairy cattle and implications for fertility. *Anim Reprod* 15(Supplement 1):879-885.

Drillich M, Beetz O, Pfutzner A, Sabin M, Sabin HJ, Kutzer H, Nattermann H and Heuwieser W. 2001. Evaluation of a systemic antibiotic treatment of toxic puerperal metritis in dairy cows. *J Dairy Sci* 84:2010-2017.

Drillich M, Raab D, Wittke M and Heuwieser W. 2005. Treatment of chronic endometritis in dairy cows with an intrauterine application of enzymes: A field trial. *Theriogenology* 63:1811-1823.

Drillich M, Tesfaye D, Rings F, Schellander K, Heuwieser W and Hoelker M. 2012. Effects of polymorphonuclear neutrophil infiltration into the endometrial environment on embryonic development in superovulated cows. *Theriogenology* 77:570-578.

Dubuc J, Duffield TF, Leslie KE, Walton JS and LeBlanc SJ. 2010a. Definitions and diagnosis of postpartum endometritis in dairy cows. *J Dairy Sci* 93:5225-5233.

Dubuc J, Duffield TF, Leslie KE, Walton JS and LeBlanc SJ. 2010b. Risk factors for postpartum uterine diseases in dairy cows. *J Dairy Sci* 93:5764-5771.

Dubuc J, Duffield TF, Leslie KE, Walton JS and LeBlanc SJ. 2011. Randomized clinical trial of antibiotic and prostaglandin treatments for uterine health and reproductive performance in dairy cows. *J Dairy Sci* 94:1325-1338.

Duchens M, Forsberg M, Edqvist LE, Gustafsson H and Rodriguez-Martinez H. 1994. Effect of induced suprabasal progesterone levels around estrus on plasma concentrations of progesterone, estradiol-17β and LH in heifers. *Theriogenology* 42:1159-1169.

Duchens M, Maciel M, Gustafsson H, Forsberg M, Rodriguez-Martinez H and Edqvist LE. 1995. Influence of peri-oestrous suprabasal progesterone levels on cycle length, oestrous behaviour and ovulation in heifers. *Anim Reprod Sci* 37:95-108.

Duffield TF, Lissemore KD, McBride BW and Leslie KE. 2009. Impact of hyperketonemia in early lactation dairy cows on health and production. *J Dairy Sci* 92:571-580.

Duncan JR, Wilkie BN, Hiestand F and Winter AJ. 1972. The serum and secretory immunoglobulins of cattle: characterisation and quantitation. *J Immunol* 108:965-976.

Duricic D, Valpotic H and Samardzija M. 2015. Prophylaxis and therapeutic potential of ozone in buiatrics: current knowledge. *Anim Reprod Sci* 159:1-7.

Đurièiæ D, Vince S, Ablondi M, Dobraniæ T and Samard•ija M. 2012. Intrauterine ozone treatment of retained fetal membrane in Simmental cows. *Anim Reprod Sci* 134:119-124.

Dzhuroua I and Gulubinov G. 1981. Histological endometrial changes in cows with latent endometritis. *Vet Med Nauki* 18:98-103.

Eckersall PD and Conner JG. 1988. Bovine and canine acute phase proteins. *Vet Res* 12:169-178.

Eckersall PD and Conner JG. 1990. Plasma haptoglobin in cattle (*Bos taurus*) exists as polymers in association with albumin. Comparative Biochemistry and Physiology Part B: *Comp Biochem* 96:309-314.

Eckersall PD, Young FJ, McComb C, Hogarth CJ, Safi S, Weber A, McDonald T, Nolan AM and Fitzpatrick JL. 2001. Acute phase proteins in serum and milk from dairy cows with clinical mastitis. *Vet Rec* 148:35-41.

Ehling-Schulz M, Svensson B, Guinebretiere MH, Lindback T, Andersson M, Schulz A, Fricker M, Christiansson A, Granum PE, Märtlbauer E, Nguyen-The C, Salkinoja-Salonen M and Scherer S. 2005. Emetic toxin formation of Bacillus cereus is restricted to a single evolutionary lineage of closely related strains. *Microbiology* 151:183-197.

El-Azab MA, Whitmore HL, Kakoma I, Brodie BO, Mckenna DJ and Gustaffson BK. 1988. Evaluation of the uterine environment in experimental and spontaneous bovine metritis. *Theriogenology* 29:1327-1334.

El-Deeb W, Elsohaby I, Fayeze M, Mkrtchyanf HV, El-Etriby D and ElGioushy M. 2020. Use of procalcitonin, neopterin, haptoglobin, serum amyloid A and proinflammatory cytokines in diagnosis and prognosis of bovine respiratory disease in feedlot calves under field conditions. *Acta Tropica* 105336.

Elliott L, McMahon KJ, Gier HT and Marion GB. 1968. Uterus of the cow after parturition: bacterial content. *Am J Vet Res* 29:77-81.

Elmore RG, Martin CE and Berg JN. 1978. Absorption of *Escherichia coli* endotoxin from the mammary glands and uteri of early postpartum sows and gilts. *Theriogenology* 10:439-446.

Elmore RG, Mollett TA, Berg JN and Schmidt DA. 1983. *Escherichia coli* endotoxin absorption from the ewe's uterus and peritoneal cavity. *Theriogenology* 19:841-847.

El-Rheem S, Ghallab R and El-Sharkawy S. 2019. SAT, a new approach in understanding and treatment of subclinical endometritis in dairy cows. *Open J Vet Med* 9:109-119.

Elstein M. 1974. The cervix and its mucus. In Clinics in Obstetrics and Gynecology 1: 345-368.

Emmanuel, DGV, Dunn SM and Ametaj BN. 2008. Feeding High Proportions of Barley Grain Stimulates an Inflammatory Response in Dairy Cows. *J Dairy Sci* 91:606-614.

Enguix A, Rey C, Concha A, Medina A, Coto D, Dieguez MA. 2001.Comparison of procalcitonin with C-reactive protein and serum amyloid for the early diagnosis of bacterial sepsis in critically ill neonates and children. *Intensive Care Med* 27:211-215.

Enkhia KL and Kohli IS. 1982. Note on the physical properties of cervico-vaginal mucus during estrus in normal and repeat breeding Rathi cows. *Indian J Anim Sci* 52:1239-1240.

Epand RF, Pollard JE, Wright JO, Savage PB and Epand RM. 2010. Depolarization, bacterial membrane composition, and the antimicrobial action of ceragenins. *Antimicrob Agents Chemother* 54:3708-3713.

Ercan N, Tuzcu N, Basbug O, Tuzcu M and Alim A. 2016. Diagnostic value of serum procalcitonin, neopterin, and gamma interferon in neonatal calves with septicemic colibacillosis. *J Vet Diagn Invest* 28:180-183.

Eriksson Å, Persson Waller K, Svennersten-Sjaunja K, Haugen JE, Lundby F and Lind O. 2005. Detection of mastitic milk using a gas-sensor array system (electronic nose). *Int Dairy J* 15:1193-1201.

Escandon BM, Espinoza JS, Perea FP, Quito F, Ochoa R, López GE, Galarza DA and Garzón JP. 2020. Intrauterine therapy with ozone reduces subclinical endometritis and improves reproductive performance in postpartum dairy cows managed in pasture-based systems. *Trop Anim Health Prod* 52:2523-2528.

Esmay PA, Billington SJ, Link MA, Songer JG and Jost BH. 2003. The *Arcanobacterium pyogenes* collagen-binding protein, CbpA, promotes adhesion to host cells. *Infect Immun* 71:4368-4374.

Espadamala A, Pereira R, Pallarés P, Lago A and Silva-Del-Río N. 2018. Metritis diagnosis and treatment practices in 45 dairy farms in California. *J Dairy Sci* 101:9608-9616.

Etherington WG, Christie KA, Walton JS, Leslie KE, Wickstrom S and Johnson WH. 1991. Progesterone profiles in postpartum Holstein dairy cows as an aid in the study of retained fetal membranes, pyometra and anestrus. *Theriogenology* 35:731-746.

Etherington WG, Martin SW, Bonnett BN, Johnson WH, Miller RB, Savage NC, Walton JS and Montgomery ME. 1988. Reproductive performance of dairy cows following treatment with cloprostenol 26 and / or 40 days postpartum: A field trial. *Theriogenology* 29:565-575.

Etzerodt A and Moestrup SK. 2013. CD163 and inflammation: biological, diagnostic, and therapeutic aspects. *Antioxid Redox Signal* 18:2352-2363.

Farin PW, Ball L, Olson JD, Mortimer RG, Jones RL, Adney WS and McChesney AE. 1989. Effect of *Actinomyces pyogenes* and gram-negative anaerobic bacteria on the development of bovine pyometra. *Theriogenology* 31:979-989.

Faundez R, Duszewska AM, Klucinski W, Spohr I and Sitarska E. 1994. Occurrence of leukocytes and epithelial cells in the lumen of the reproductive tract during the ovarian cycle. In: Proceedings of the 18th World Buiatr. Congress. Bologna, Italy, pp. 305-308.

Fecteau KA, Eiler H, 1996: Evaluation of injections of collagenase and oxytetracycline via the umbilical artery as treatment for retained placenta in cattle. *Am J Vet Res* 57:522-525.

Fernandez Olmos A, García de la Fuente C, Saénz Nieto J, Valdezate Ramos S. 2010. Métodos de Identificación Bacteriana en el Laboratorio de Microbiología. *Procedimientos de Microbiología Clínica* 52.

Ferris RA. 2021. Bacterial film and endometritis. Chapter 37. In: *Equine Reproductive Procedures*. John Dascanio and Patrick McCue (Eds). https://doi.org/10.1002/9781119556015.ch37

Fischer C, Drillich M, Odau S, Heuwieser W, Einspanier R and Gabler C. 2010. Selected pro-inflammatory factor transcripts in bovine endometrial epithelial cells are regulated during the oestrous cycle and elevated in case of subclinical or clinical endometritis. *Reprod Fertil Dev* 22:818-829.

Fishman N. 2006. Antimicrobial stewardship. *Am J Infect Contr* 34:S55–63; discussion S64-73.

Flores J, Dupont HL, Jiang ZD, Okhuysen PC, Melendez-Romero JH, Gonzalez-Estrada A, Carrillo I and Paredes M. 2011. A randomized, double-blind, pilot study of rifaximin 550 mg versus placebo in the prevention of travelers' diarrhea in Mexico during the dry season. *J Travel Med* 18:333-336.

Földi J, Kulcsár M, Pécsi A, Huyghe B, de Sa C, Lohuis JA, Cox P and Huszenicza G. 2006 Bacterial complications of postpartum uterine involution in cattle. *Anim Reprod Sci* 96:265-281.

Fonseca FA, Britt JH, McDaniel BT, Wilk JC and Rakes AH. 1983. Reproductive traits of Holsteins and Jerseys. Effects of age, milk yield, and clinical abnormalities on involution of cervix and uterus, ovulation, estrous cycles, detection of estrus, conception rate and days open. *J Dairy Sci* 66:1128-1147.

Fox A, Lee CS, Brandon MR and Meeusen ENT. 1998. Effects of pregnancy on lymphocytes within sheep uterine interplacentomal epithelium. *Am J Reprod Immunol* 40:295-302.

Franasiak JM, Werner MD, Juneau CR, Tao X, Landis J, Zhan Y, Treff NR and Scott RT 2016 Endometrial microbiome at the time of embryo transfer: next-generation sequencing of the 16s ribosomal subunit. *J Assist Reprod Genet* 33:129-136.

Frank T, Anderson KL, Smith AR, Whitmore HL and Gustafsson BK. 1983. Phagocytosis in the uterus: a review. *Theriogenology* 20:103-111.

Fray MD, Mann GE, Clarke MC and Charleston B. 1999. Bovine viral diarrhea virus: its effects on estradiol, progesterone and prostaglandin secretion in the cow. *Theriogenology* 51:1533-1546.

Frazer GS, Rossol TJ, Threlfall WR and Weisbrode SE. 1988. Histopathologic effects of dimethyl sulfoxide on equine endometrium. *Am J Vet Res* 49:1774-1781.

Fredriksson G, Kindahl H, Sandstedt K and Edqvist LE. 1985. Intrauterine bacterial findings and release of PGF2_ in the postpartum dairy cow. *Zbl Vet Med A* 32:368-380.

Fredriksson G. 1984. Some reproductive and clinical aspects of endotoxin in cows with special emphasis on the role of prostaglandins. *Acta Vet Scan* 25:365-377.

Freeman M, Ashkenas J, Rees DJ, Kingsley DM, Copeland NG, Jenkins NA and Krieger M. 1990. An ancient, highly conserved family of cysteine-rich protein domains revealed by cloning type I and type II murine macrophage scavenger receptors. *Proc Natl Acad Sci* USA, 87:8810-8814.

Fricke PM. 2002. Scanning the future ultrasonography as reproductive management tool for dairy cows. *J Dairy Sci* 85:1918-1926.

Froom P, Bieganiec B, Ehrenrich Z and Barak M. 2000. Stability of common analytes in urine refrigerated for 24 h before automated analysis by test strips. *Clin Chem* 46:1384-1386.

Fuentes BM, Arias LAQ, Gonzalez JJB, del Sol LDB, Feijoo JEM, Punal JLG, Lopez, MB, Lago AP, Cao JMD, Rodriguez GF, Herradon PJG and Martinez AIP. 2017. Agreement between postmortem endometrial cytology, biopsy and bacteriology in culled dairy cows. *Anim Reprod* 14:1024-1033.

Fukushima Y. 1972. A etiological factors concerned E. coli infections of the bovine female genital system. *Bull Azabu Vet Col* 23:123-160.

Fuquay JW, Harris RA, McGee WH, Bcalty JA and Arnold BL. 1975. Routine postpartum treatment of dairy cows with intrauterine neomycin sulfate boluses. *J Dairy Sci* 58:1367-1369.

Gabler C, Fischer C, Drillich M, Einspanier R and Heuwieser W. 2010. Time-dependent mRNA expression of selected pro-inflammatory factors in the endometrium of primiparous cows postpartum. *Reprod Biol Endocrinol* 8:152.

Gabler C, Drillich M, Fischer C, Holder C, Heuwieser W and Einspanier R. 2009. Endometrial expression of selected transcripts involved in prostaglandin synthesis in cows with endometritis. *Theriogenology* 71:993-1004.

Galvao KN, Felippe MJ, Brittin SB, Sper R, Fraga M, Galvão JS, Caixeta L, Guard CL, Ricci A and Gilbert RO. 2012. Evaluation of cytokine expression by blood monocytes of lactating Holstein cows with or without postpartum uterine disease. *Theriogenology* 77:356-372.

Galvao KN, Flaminio MJBF, Brittin SZ, Sper R, Fraga M, Caixeta L, Ricci A, Guard CL, Butler WR and Gilbert RO. 2010 b. Association between uterine disease and indicators of neutrophil and systemic energy status in lactating Holstein cows. *J Dairy Sci* 93:2926-2937.

Galvão KN, Frajblat M, Brittin SB, Butler WR, Guard CL, and Gilbert RO. 2009b. Effect of prostaglandin F2alpha on subclinical endometritis and fertility in dairy cows. *J Dairy Sci* 92:4906-4913.

Galvao KN, Frajblat M, Butler WR, Brittin SB, Guard CL and Gilbert RO. 2010a. Effect of early postpartum ovulation in dairy cows. *Reprod Dom Anim* 45:207-211.

Galvão KN, Greco LF, Vilela JM, Sá Filho MF and Santos JEP. 2009a. Effect of intrauterine infusion of Ceftiofur on uterine health and fertility in dairy cows. *J Dairy Sci* 92:1532-1542.

Galvao KN, Santos NR, Galvão JS and Gilbert RO. 2011. Association between endometritis and endometrial cytokine expression in postpartum Holstein cows. *Theriogenology* 15:76:290-299.

Galvao KN. 2011. Identifying and Treating Uterine Disease in Dairy Cows. *Proceedings 47th Florida Dairy Production Conference,* Gainesville, March 30, 2011:21-29.

Galvao KN. 2018. Association between immune function and development of uterine disease in dairy cows. *Anim Reprod* 9:318-322.

Ganz T, Selsted ME, Szklarek D, Harwig SSL, Daher K, Bainton DF and Lehrer RI. 1985. Defensins. Natural peptide antibiotics of human neutrophils. *J Clin Invest* 76:1427-1435.

Garcia-Villar R, Marnet PG, Laurentie MP, Toutain PL. 1987. Fenprostalene in cattle: Evaluation of oxytocic effects in ovariectomized cows and abortion potential in a 100-day pregnant cow. *Theriogenology* 28:467-480.

Gartner MA, Bondzio A, Braun N, Jung M, Einspanier R and Gabler C. 2015. Detection and characterisation of Lactobacillus spp. in the bovine uterus and their influence on bovine endometrial epithelial cells *in vitro*. *PLoS One* 10: p 0119793.

Gatne MM, Talong AG, Toke SK, Kalkar RD and Ranade VV. 1993. Pharmcokinetics of neomycin in calves. *Indian Vet J* 70:422-425.

Gatta L and Scarpignato C. 2017. Systematic review with meta-analysis: rifaximin is effective and safe for the treatment of small intestine bacterial overgrowth. *Aliment Pharmacol Ther* 45:604-616.

Gautam G, Nakao T, Yusuf M, Koike K. 2009. Prevalence of endometritis during the postpartum period and its impact on subsequent reproductive performance in two Japanese dairy herds. *Anim Reprod Sci* 116:175-187.

Gawronska A, Banasiuk M, Lachowicz D, Pituch H, Albrecht P, Banaszkiewicz A. 2017. Metronidazole or rifaximin for treatment of *Clostridium difficile* in pediatric patients with inflammatory bowel disease: A randomized clinical trial. *Inflamm Bowel Dis* 23:2209-2214.

Gay JM and Upham GL. 1994. Effect of exogenous prostaglandin F2α in clinically normal postparturient dairy cows with a palpable corpus luteum. *J Am Vet Med Assoc* 205:870-873.

Geary TW, Smith MF, MacNeil MD, Day ML, Bridges GA, Perry GA, Abreu FM, Atkins JA, Pohler KG, Jinks EM and Madsen CA. 2013. Triennial Reproduction Symposium: influence of follicular characteristics at ovulation on early embryonic survival. *Jf Anim Sci* 91:3014-3021.

Gennaro R, Scheidder C, DE Nicola G, Cian F and Romeo D. 1978. Biochemical properties of bovine Granulocytes. *Proc Soc Exp Bio Med* 157:342-347.

Georgiv S. 1978. Study of the general immunologic response and microbial content of the uterus of cows in the puerperal period. *Vet Sci* 15:69-77.

Ghallab RS, Gad El Karim DRS, Abdel Hasseb Fayed AH and Rashad AMA. 2023. Efficiency of conventional and nanoparticle oxytetracycline in treatment of clinical endometritis in postpartum dairy cows. *Trop Anim Health Prod* 55:118

Ghanem ME, Tezuka E, Devkota B, Izaike Y and Osawa T. 2015. Persistence of uterine bacterial infection, and its associations with endometritis and ovarian function in postpartum dairy cows. *J Reprod Dev* 61:54-60.

Ghanem ME, Tezuka E, Sasaki K, Takahashi M, Yamagishi N, Izaike Y and Osawa T. 2016. Correlation of blood metabolite concentrations and body condition scores with persistent postpartum uterine bacterial infection in dairy cows. *J Reprod Dev* 62:457-463.

Ghasemi F, Gonzalez-Cano P, Griebel PJ and Palmer C. 2012. Proinflammatory cytokine gene expression in endometrial cytobrush samples harvested from cows with and without subclinical endometritis. *Theriogenology* 78:1538-1547.

Giannattasio-Ferraz S, Laguardia-Nascimento M, Gasparini MR, Leite LR, Araujo FMG, de Matos Salim AC, de Oliveira AP, Nicoli JR, de Oliveira GC, da Fonseca FG and Barbosa-Stancioli EF. 2019. A common vaginal microbiota composition among breeds of Bos taurus indicus (Gyr and Nellore). *Braz J Microbiol* 50:1115-1124.

Giannopoulos G, Jackson K, Kredentser J and Tulchinsky D. 1985. Prostaglandin E and F2α receptors in human myometrium during the menstrual cycle and in pregnancy and labor. *Am J Obst Gynae* 153:904-910.

Gier HT and Marion GB. 1968. Uterus of the cow after parturition: involutional changes. *Am J Vet Res* 29: 83-96.

Gilbert RO and Santos NR. 2016. Dynamics of postpartum endometrial cytology and bacteriology and their relationship to fertility in dairy cows. *Theriogenology* 85:1367-1374.

Gilbert RO, Bosu WTK and Peter AT. 1990. The effect of *Escherichia coli* endotoxin on luteal function in Holstein heifers. *Theriogenology 33*:645-651.

Gilbert RO, Shin ST, Guard CL, Erb HN and Frajblat M. 2005. Prevalence of endometritis and its effects on reproductive performance of dairy cows. *Theriogenology* 64:1879-1888.

Gilbert RO, Grohn YT, Miller PM and Hoffman DJ. 1993. Effect of parity on periparturient neutrophil function in dairy cows. *Vet Immunol Immunopathol* 36:75-82.

Gilbert RO. 1992. Bovine endometriris: The burden of proof. *Cornett Vet* 82:11-14.

Gill RS, Gangwar PC and Koomer DS. 1973. Optimum time of insemination for high conception rate in buffaloes. *Indian J Anim Sci* 43:355-357.

Glanvill S and Dobson H. 1991. Effect of prostaglandin treatment on the fertility of problem cows. *Vet Rec* 128:374-376.

Glasel JA. 1995. Validity of nucleic acid purities monitored by 260nm/280nm absorbance ratios. *Biotechniques* 18:62-63.

Glenthoj A, Bostofle E and Rank F. 1986. Brush cytology from the uterine endocervix. *Acta Obstet Gynecol Scand* 65:689-691.

Goff JP and Horst RL. 1997. Physiological changes at parturition and their relationship to metabolic disorders. *J Dairy Sci* 80:1260-1268

Goff JP. 1999. Treatment of calcium, phosphorus, and magnesium balance disorders. *Vet Clin North Am Food Anim Prac* 15:619-639.

Gohil P, Patel K, Purohit K, Chavda D, Puvar A, Suthar VS, Patil DB, Joshi M and Joshi CG. 2022. Metagenomic analysis of bacterial communities during estrous cycle in *Bos indicus*. *Trop Anim Health Prod* 54:1-11.

Gonzales HE, Crowl WA, Caudle AB and Thompson FN. 1985. Morphometric studies of the bovine uterus: Microscopic lesions and retrospective reproductive history. *Am J Vet Res* 46:2588-2595.

Gorden PJ, Ydstie JA, Kleinhenz, MD, Wulf LW, Gehring R, Lee CJ, Wang C and Coetzee JF. 2016. A study to examine the relationship between metritis severity and depletion of oxytetracycline in plasma and milk after intrauterine infusion. *J Dairy Sci* 99:8314-8322.

Gores-Lindholm A, Ahlschwede S, Causey R, Calderwoods-Mays M and Le Blanc MM. 2009. Effect of intra-uterine infusion of diluted N-acetylcysteine on equine endometrium. *AAEP Proc*, 55, 326.

Gorzecka J, Callesen H, Pedersen KM and Friggens NC. 2011. The relationship between postpartum vaginal discharge symptoms and progesterone profile characteristics in lactating dairy cows in Denmark. *Theriogenology* 75:1016-1028.

Gottshall SL and Hansen PJ. 1992. Regulation of leucocyte subpopulations in sheep endometrium by progesterone. *Immunology* 76:636-641.

Granovsky-Grisaru S, Ilan D, Grisaru D, Lavie O, Aboulafia I, Diamant YZ and Hanani M. 1998. Effects of erythromycin on contractility of isolated myometrium from pregnant rats. *Am J Obstet Gynecol* 178:171-174.

Griffin JFT, Hartigan PJ and Nunn WR. 1974. Non-specific uterine infection and bovine fertility: I. Infection patterns and endometritis during the first seven weeks post-partum. *Theriogenology* 1:91-106.

Grohn YT, Erb HN, McCulloch CE and Saloniemi HS. 1990. Epidemiology of reproductive disorders in dairy cattle: association among host characteristics, disease and production. *Prev Vet Med* 8:25-39.

Gronlund U, Sandgren CH and Waller KP. 2005. Haptoglobin and serum amyloid A in milk from dairy cows with chronic sub-clinical mastitis. *Vet Res* 36:191-198.

Grunert E. 1979. Zur Problematik der rektalen Ovarkontrolle beim Rind Prakt Tierarzt 60:13-18.

Grunert T, Wenning M, Barbagelata MS, Fricker M, Sordelli DO, Buzzola FR and Ehling-Schulz M. 2013. Rapid and reliable identification of *Staphylococcus aureus* capsular serotypes by means of artificial neural network-assisted Fourier transform infrared spectroscopy. *J Clin Microbiol* 51:2261-2266.

Gruys E, Ederen AM, Alsemgeest SPM, Kalsbeek HC and Wensing T. 1993. Acute phase protein values in blood of cattle as indicator of animals with pathological process. *Arch Lebensmittel Hyg* 44:107-111.

Guedawy SA, Neff-Davis CA, Davis LE *et al.* 1983b. Disposition of gentamicin in the genital tract of cows. *J Vet Pharmacol Therap* 6:85-92.

Guedawy SA, Yasquez L, Neff-Davis *et al.* 1983a. Effect of vehicle on intrauterine absorption of gentamicin in cattle. *Theriogenology* 19:771-778.

Guidry AJ, Paapa MJ and Pearson RE. 1976. Effects of parturition and lactation on blood and milk concentrations, corticosteroids and neutrophil phagocytosis in the cow. *Am J Vet Res* 37:1195-1200.

Guilbault LA, Thatcher WW, Drost M and Hopkins SM, 1984a. Source of F series prostaglandins during the early postpartum period in cattle. *Biol Reprod* 31:879-887.

Guilbault LA, Thatcher WW, Gross M and Haibel GK. 1987. Influence of physiological infusion of prostaglandin F2α into postpartum cows with partially suppressed endogenous production of prostaglandins. 1. Uterine and ovarian morphological responses. *Theriogenology* 27:931-946.

Guilbaut LA, Thatcher WW, Foster DB and Caton D. 1984b. Relationship of 15-keto-13,14-dihydroprostaglandin F2α concentrations in peripheral plasma with local uterine production of F series prostaglandins and changes in uterine blood flow during the early postpartum period of cattle. *Biol Reprod* 31:870-878

Gunnink JW. 1973. Een Onderzoek naar het Afweermechanisme Van De Uterus. University of Utrecht, Ph.D. Thesis, pp. 143-160.

Gupta JP, Shyma KP, Mohandas AC and Sneharaj RK. 2011. White side test to study subclinical endometritis in crossbred cattle. *Bhartya Krishi Anushandhan Patrika* 26:123-125.

Gupta RC, Sinha AK and Krishnaswamy A. 1983. Studies on the efficacy of some post-service intrauterine infusions on the conception rate in repeat breeding cattle. *Theriogenology* 20:559-564.

Gupta SK, Prakash J and Srivastava S. 2002. Validation of traditional claim of Tulsi, Ocimum sanctum Linn as a medicinal plant. *Indian J Exp Biol* 40:765-773.

Gurunathan S, Choi YJ and Kim JH. 2018. Antibacterial efficacy of silver nanoparticles on endometritis caused by *Prevotella melaninogenica* and *Arcanobacterum pyogenes* in dairy cattle. *Int J Mol Sci* 19:1210.

Gurunathan S, Han JW, Kwon DN and Kim JH. 2014b. Enhanced antibacterial and anti-biofilm activities of silver nanoparticles against gram-negative and gram-positive bacteria. *Nanoscale Res Lett* 9:373.

Gurunathan S, Jeong JK, Han JW, Zhang XF, Park JH and Kim JH. 2015. Multidimensional effects of biologically synthesized silver nanoparticles in *Helicobacter pylori*, *Helicobacter felis*, and human lung (L132) and lung carcinoma A549 cells. *Nanoscale Res Lett* 10:35.

Gurunathan S. Han J, Park JH and Kim JH. 2014a. A green chemistry approach for synthesizing biocompatible gold nanoparticles. *Nanoscale Res Lett* 9:248.

Gurunathan, S. 2015. Biologically synthesized silver nanoparticles enhances antibiotic activity against gram-negative bacteria. *J Ind Eng Chem* 29:217-226.

Guslandi M, Petrone MC and Testoni PA. 2006. Rifaximin for active ulcerative colitis. *Inflamm Bowel Dis* 12:335.

Gustafasson BK. 1980. Disposition of antimicrobial drugs in the female genital system. *Proceedings 9th ICAR*. 2:485-490.

Gustafsson BK and Ott RS. 1981. Current trends in the treatment of genital infections in large animals. *Comp Cont Education* 5:147-151.

Gustafsson BK. 1984. Therapeutic strategies involving anti-microbial treatment of the uterus of large animals. *J Am Vet Med Assoc* 185:1194-1198.

Gustafsson H, Larsson K, Kindahl H and Madej A. 1986. Sequential endocrine changes and behaviour during oestrus and metoestrus in repeat breeder and virgin heifers. *Anim Reprod Sci* 10:261-273.

Guven F M, Aydin H, Yildiz G, Engin A, Celik VK, Bakir D and Deveci K. 2013. The importance of myeloperoxidase enzyme activity in the pathogenesis of Crimean-Congo haemorrhagic fever. *J Med Microbiol* 62:441-445.

Guyton, A.C., 1991. The lymphatic system. Textbook of Medical Physiology, 8th Edition. Saunders, Philadelphia, pp. 180-184.

Güzel-Seydim Z, Bever PI and Greene AK. 2004. Efficacy of ozone to reduce bacterial populations in the presence of food components. *Food Microbiol* 21:475-479.

Gymnich S, Hiss S, Sauerwein H and Petersen B. 2003. Haptoglobin in sows at parturition. *4th European Colloquium on Acute Phase Proteins*. Segovia, Spain.

Haddad NS, Pedersoli WM, Ravis WR, Fazeli MH and Carson Jr RL. 1985. Pharmacokinetics of gentamicin at steady state in ponies: Serum, urine and endometrial concentrations. *Am J Vet Res* 46:1268-1271.

Haddad NS, Ravis WR, Pedersoli WM and Carson Jr RL. 1987. Pharmacokinetics and tissue residues of gentamicin in lactating cows after multiple intramuscular doses are administered. *Am J Vet Res* 48:21-27.

Hafez ESE and Hafez B. 2000. Reproduction in farm animals. Lippincott Williams and Wilkins, Baltimore, Maryland, USA.

Haimerl P and Heuwieser W. 2014. Invited review: Antibiotic treatment of metritis in dairy cows: a systematic approach. *J Dairy Sci* 97:6649-6661.

Haimerl P, Arlt S, Borchardt S and Heuwieser W. 2017. Antibiotic treatment of metritis in dairy cows—A meta-analysis. *J Dairy Sci* 100:3783-3795.

Hajurka J, Nagy J, Popelka P, Rozanska H, Sokol J, Cabadaj R and Hura V. 2003. Tetracycline concentrations in blood and milk of cows following intrauterine treatment of acute or subacute/chronic endometritis. *Bull Vet Inst Pulawy* 47:435-447.

Hamana K, El-Banne AA, Hafez ESE. 1971. Sialic acid and some physiological characteristic of bovine cervical mucus. *Cornell Vet* 61:104-111.

Hammon DS, Evjen IM, Dhiman TR, Goff JP and Walters JL. 2006. Neutrophil function and energy status in Holstein cows with uterine health disorders. *Vet Immunol Immunopath* 113:21-29.

Hartigan PJ, Murphy JA, Nunn WR and Griffin JFT. 1972. investigation into the causes of reproductive failure in dairy cows. II. Uterine infection and endometrial histopathology in clinically normal repeat-breeder cows. *Irish Vet J* 26:245-247.

Haskell SR, Gehring R, Payne MA, Craigmill AL, Webb AI, Baynes RE and Riviere JE. 2003. Update on FARAD food animal drug withholding recom-mendations. *J Am Vet Med Assoc* 223:1277-1278

Hauser WE, JR MD and Remington JS. 1982. Effect of antibiotics on the immune response. *Am J Med* 72:711-716.

Hawk HW, Brinfield TH, Turner GD, Whitmore GW and Norcross MA. 1964. Effect of ovarian status on induced and inflammatory response in cattle uteri. *Am J Vet Res* 25:362-366.

Hawk HW, Turner GD and Sykes J. 1960. The effect of ovarian hormones on the uterine defense mechanism during the early stages of induced infection. *Am J Vet Res* 21:644-656.

Hawk HW. 1983. Sperm survival and transport in the female reproductive tract. *J Dairy Sci* 66:2645-2660.

Hayashi A, Shinohara M and Ohura K. 1999. Effect of insulin on naturally occurring gingivitis rats with diabetes. *J Osaka Dent Univ* 33:1-7.

Heidari M, Kafi M, Mirzaei A, Asaadi A and Mokhtari A. 2019. Effects of follicular fluid of preovulatory follicles of repeat breeder dairy cows with subclinical endometritis on oocyte developmental competence. *Anim Reprod Sci* 205:62-69.

Heidarpour M, Mohri M, Fallah Rad AH, Shahreza FD and Mohammadi M. 2012. Acute-phase protein concentration and metabolic status affect the outcome of treatment in cows with clinical and subclinical endometritis. *Vet Rec* 171:1-5.

Heiser A, LeBlanc SJ and McDougall S. 2018. Pegbovigrastim treatment affects gene expression in neutrophils of pasture-fed, periparturient cows. *J Dairy Sci* 101:8194-8207.

Herath S, Dobson H, Bryant CE and Sheldon IM. 2006. Use of the cow as a large animal model of uterine infection and immunity. *J Reprod Immunol* 69:13-22.

Herath S, Lilly ST, Santos NR, Gilbert RO, Goetze L, Bryant CE, White JO, Cronin J and Sheldon IM. 2009. Expression of genes associated with immunity in the endometrium of cattle with disparate postpartum uterine disease and fertility. *Reprod Biol Endocrinol* 7:1-13.

Herath S, Williams EJ, Lilly ST, Gilbert RO, Dobson H, Bryant CE and Sheldon IM. 2007. Ovarian follicular cells have innate immune capabilities that modulate their endocrine function. *Reproduction* 134:683-693.

Herath S, Lilly ST, Fischer DP, Williams EJ, Dobson H, Bryant CE and Sheldon IM. 2009. Bacterial lipopolysaccharide induces an endocrine switch from prostaglandin F2alpha to prostaglandin E2 in bovine endometrium. *Endocrinology* 150:1912-1920.

Herdt TH. 2000. Ruminant adaptation to negative energy balance. *Vet Clin North Am Food Anim Pract* 16:215-230.

Hernandez EGS, Bouda J, Cecilio AAA, Doubek J and Forero FHV. 2014. Effect of prostaglandin F2α administration during the first postpartum hours on the serum calcium concentration in dairy cows. *Veterinaria México OA* vol.1 no.2 Ciudad de México oct./dic. 2014.

Herzog K, Debertolis L, Kastelic JP, Schmicke M, Ulbrich SE and Bollwein H. 2019. Effects of intravenous infusion of *E. coli* lipopolysaccharide in early pregnant cows. *Reproduction* 157:65-76.

Heuwieser W, Tenhagen BA, Tischer M, Lu¨hr J and Blum H. 2000. Effect of three programs for the treatment of endometritis on the reproductive performance of a dairy herd. *Vet Rec* 146:338-341.

Hill EK and Li J. 2017. Current and future prospects for nanotechnology in animal production. *J Anim Sci Biotechnol* 8:26.

Hirsbrunner G, Burkhardt HW and Steiner A. 2006. Effects of a single administration of prostaglandin F2alpha, or a combination of prostaglandin F2alpha and prostaglandin E2, or placebo on fertility variables in dairy cows 3–5 weeks post partum, a randomized, double-blind clinical trial. *Reprod Biol Endocrinol* 4:65.

Hirsbrunner G, Knutti B, Kupfer U, Burkhardt H, Steiner A. 2003. Effect of prostaglandin E2, DL-cloprostenol, and prostaglandin E2 in combination with D-cloprostenol on uterine motility during diestrus in experimental cows. *Anim Reprod Sci* 79:17-32.

Hirsbrunner G, Küpfer U, Burkhardt H and Steiner A. 1998. Effect of different prostaglandins on intrauterine pressure and uterine motility during diestrus in experimental cows. *Theriogenology* 50:445-455.

Hirvonen J and Pyorala S. 1998. Acute-phase response in dairy cows with surgically-treated abdominal disorders. *Vet J* 155:53-61.

Hirvonen J, Huszenicza G, Kulcsar M and Pyorala S. 1999. Acute-phase response in dairy cows with acute postpartum metritis. *Theriogenology* 51:1071-1083.

Hisaeda K, Hagiwara K, Eguchi J, Yamanaka H, Kirisawa R and Iwai H. 2001. Interferon-ã and tumor necrosis factor-α levels in sera and whey of cattle with naturally occurring coliform mastitis. *J Vet Med Sci* 63:1009-1011.

Hisatsune K, Nozaki S, Ishikawa T, Hayashi M, Nogaki K and Ogawa H. 1983. A biochemical study of the phagocytic activities of tuftsin and its analogues. *Annals N Y Acad Sci* 419:205-213.

Hiss S, Mielenz M, Bruckmaier R M and Sauerwein H. 2004. Haptoglobin Concentrations in Blood and Milk After Endotoxin Challenge and Quantification of Mammary Hp mRNA Expression. *J Dairy Sci* 87:3778-3784.

Hiss S, Weinkauf C, Hachenberg S and Sauerwein H. 2009. Short communication: Relationship between metabolic status and the milk concentrations of haptoglobin and lactoferrin in dairy cows during early lactation. *J Dairy Sci* 92:4439-4443.

Hoedemaker M, Lund LA and Wagner WC. 1992. Influence of arachidonic acid metabolites and steroids on function of bovine polymorphonuclear neutrophils. *Am J Vet Res* 53:1534-1543.

Hoelker M, Salilew-Wondim D, Drillich M, Christine G, Ghanem N, Goetze L, Tesfaye D, Schellander K and Heuwieser W. 2012. Transcriptional response of the bovine endometrium and embryo to endometrial polymorphonuclear neutrophil infiltration as an indicator of subclinical inflammation of the uterine environment. *Reprod Fertil Dev* 24:778-793.

Hoeprich PD and Martin CH. 1989. Effect of tetracycline, polymyxin B, and rifampin on phagocytosis. *Clin Pharmacol Ther* 11:418-422.

Hoffman JT, Hartig C, Sonbol E and Lang M. 2011. Probable interaction between warfarin and rifaximin in a patient treated for small intestine bacterial overgrowth. *Ann Pharmacother* 45:e25.

Hofmann GE, Rao ChV, Barrows GH and Sanfilippo JS. 1983. Topography of human uterine prostaglandin E and F2α Receptors and their profiles during pathological states. *J Clinical Endocrinol Metab* 57:360-366.

Hopkins FM. 1983. Prostaglandins and postpartum bovine uterus. In: *Proceedings of the Annual Meeting of Society for Theriogenology*. Nashville, pp. 124-128.

Hopper RM. 2014. Bovine Reproduction. John Wiley and Sons. Chapter 74: P Vincent, SL Underwood, C Dolbec, N Bouchard, T Kroetsch and P Blondin. 685-695.

Horadagoda A, Eckersall PD, Hodgson JC, Gibbs HA and Moon GM. 1994. Immediate responses in serum TNF-α and acute phase protein concentrations to infection with Pasteurella haemolytica A1 in calves. *Res Vet Sci* 57:129-132.

Horadagoda N U, Knox KMG, Gibbs HA, Reid SWJ, Horadagoda A, Edwards SER and Eckersall PD. 1999. Acute phase proteins in cattle: discrimination between acute and chronic inflammation. *Vet Rec* 144:437-441.

Hu Y, Ren J, Zhan M, Li W and Dai H. 2012. Efficacy of rifaximin in prevention of travelers' diarrhea: a meta analysis of randomized, double blind, placebo controlled trials. *J Travel Med* 19:352-356.

Huang J, Hawkinson D and Miller GE. 1972. Passage of sulfamethoxine from the bovine uterus into blood, milk and urine. *J Dairy Sci* 55:705-711.

Hudson M and Schuchmann M. 2019. Long-term management of hepatic encephalopathy with lactulose and/or rifaximin: a review of the evidence. *Eur J Gastroenterol Hepatol* 31:434-450.

Hughes JP and Couto MA. 1988. Endometritis in the mare. In: Proceedings of the 11th International Congress on Animal Reproduction and AI, Vol. 5. Dublin, pp. 91-99.

Humblet MF, Guyot H, Boudry B, Mbayahi F, Hanzen C, Rollin F and Godeau JM. 2006. Relationship between haptoglobin, serum amyloid A, and clinical status in a survey of dairy herds during a 6-month period. *Vet Clin Pathol* 35:188-193.

Hussain AM and Daniel RCW. 1992. Effects of intrauterine infusion of *Escherichia coli* endotoxin in normal cows and in cows with endometritis induced by experimental infection with *Streptococcus agalactia*. *Theriogenology* 37:791-810.

Hussain AM, Daniel RCW and O'Boyle D. 1990. Postpartum uterine flora following normal and abnormal puerperium in cows. *Theriogenology* 34:291-302.

Hussein EK and Hussein KA. 2022. Assess the efficiency of silver nanoparticles for treatment of endometritis in Iraqi breed cows. *Egypt J Vet Sci* 53:263-271.

Huszenicza G, Fodor M, Gacs M, Kulcsar M, Dohmen MJW, Vamos M, Porkolab L, Kegl T, Bartyik J, Lohuis JACM and Janosi S. 1999. Uterine bacteriology, resumption of cyclic ovarian activity and fertility in postpartum cows kept in large-scale dairy herds. *Reprod Dom Anim* 34:237-245.

Huzzey JM, Duffield TF, LeBlanc SJ, Veira DM, Weary DM and von Keyserlingk MA. 2009. Short communication: Haptoglobin as an early indicator of metritis. *J Dairy Sci* 92:621-625.

Huzzey JM, Mann S, Nydam DV, Grant R J and Overton TR. 2015. Associations of peripartum markers of stress and inflammation with milk yield and reproductive performance in Holstein dairy cows. *Prev Vet Med* 120:291-297.

Huzzey JM, Nydam DV, Grant RJ and Overton TR. 2011. Associations of prepartum plasma cortisol, haptoglobin, fecal cortisol metabolites, and nonesterified fatty acids with postpartum health status in Holstein dairy cows. *J Dairy Sci* 94:5878-5889.

Huzzey JM, Veira DM, Weary DM and von Keyserlingk MAG. 2007. Prepartum behaviour and dry matter intake identify dairy cows at risk for metritis. *J Dairy Sci* 90:3220-3233.

Hyde B and Brennan M. 2020. Does intrauterine cephapirin improve subsequent fertility in cows with clinical endometritis? *Vet Rec* 131:350-352.

Inoue H, Shinohara M and Ohura K. 1997. The effect of leukocyte function of streptozotocin induced diabetes in naturally occurring gingivitis rat. *J Osaka Dent Univ* 31:47–54.

Ishikawa Y, Nakada K, Hagiwara K, Kirisawa R, Iwai H, Moriyoshi M and Sawamukai Y. 2004. Changes in interleukin-6 concentration in peripheral blood of pre-and post-partum dairy cattle and its relationship to postpartum reproductive diseases. *J Vet Med Sci* 66:1403-1408.

Islama R, Kumar H, Nandi S and Rai RB. 2013. Determination of anti-inflammatory cytokine in periparturient cows for prediction of postpartum reproductive diseases. *Theriogenology* 79:974-979.

Janeway Jr CA, Travers P, Walport M and Shlomchik MJ. 2001. Infectious agents and how they cause disease. In Immunobiology: *The Immune System in Health and Disease*. 5th edition. Garland Science.

Jaureguiberry M, Giuliodori MJ, Mang AV, Madoz LV, Pothmann H, Drillich M and de la Sota RL. 2017. Repeat breeder cows with fluid in the uterine lumen had poorer fertility. *J Dairy Sci* 100:3083-3085.

Jaureguiberry M, Madoz LV, Giuliodori MJ, Wagener K, Prunner I, Grunert T, Ehling-Schulz M, Drillich M and de la Sota RL. 2016. Identification of *Escherichia coli* and *Trueperella pyogenes* isolated from the uterus of dairy cows using routine bacteriological testing and Fourier transform infrared spectroscopy. *Acta Vet Scand.* 28:58(1):81.

Javed MT and Khan MZ. 1991. Bacteriological and biohistopathological studies in repeat breeding cows. *J Islamic Acad Sci* 4:242-244.

Jayachandran C, Singh MK and Banerjee MC. 1990. Pharmacokinetics and distribution of ampicillin in plasma, milk and uterine fluid of female buffaloes. *Vet Res Commun* 14:47-51.

Jayachandran C, Singh MK, Singh SD and Banerjee MC. 1988. Pharmacokinetics of streptomycin with particular reference to its distribution in plasma, milk and uterine fluid of she buffaloes. *Vet Res Commun* 11:355-358.

Jayalakshmi K, Paramasivam M, Sasikala M, Tamilam TV, Sumithra A. 2017. Review on antibiotic residues in animal products and its impact on environments and human health. *J Entomol Zool Stud* 5:1446-1451.

Jayappa GH and Loken KI. 1983. Effect of antimicrobial agents and corticosteroids on bovine polymorphonuclear leukocyte chemotaxis. *Am J Vet Res* 44:2155-2159.

Jeon SJ and Galvao KN. 2018. An advanced understanding of uterine microbial ecology associated with metritis in dairy cows. *Genom Inform* 16(4) :e21.

Jeon SJ, Cunha F, Ma X, Martinez N, Vieira-Neto A, Daetz R, Bicalho RC, Lima S, Santos JEP, Jeong KC and Galvao KN. 2016. Uterine microbiota and immune parameters associated with fever in dairy cows with metritis. *PLoS One* 11: e0165740.

Jeon SJ, Lima FS, Vieira-Neto A, Machado VS, Lima SF, Bicalho RC, Santos JEP and Galvão KN. 2018. Shift of uterine microbiota associated with antibiotic treatment and cure of metritis in dairy cows. *Vet Microbiol* 214:132-139.

Jeon SJ, Vieira-Neto A, Gobikrushanth M, Daetz R, Mingoti RD, Parize AC, de Freitas SL, da Costa AN, Bicalho RC, Lima S, Jeong KC and Galvão KN. 2015. Uterine microbiota progression from calving until establishment of metritis in dairy cows. *Appl Environ Microbiol* 81:6324-6332.

Jeremejeva J, Orro T, Waldmann A and Kask K. 2012. Treatment of dairy cows with PGF2α or NSAID, in combination with antibiotics, in cases of postpartum uterine inflammation. *Acta Vet Scand* 54:45

Jinks EM, Smith MF, Atkins JA, Pohler KG, Perry GA, MacNeil MD, Roberts AJ, Waterman RC, Alexander LJ and Geary TW. 2013. Preovulatory estradiol and the establishment and maintenance of pregnancy in suckled beef cows. *J Anim Sci* 91:1176-1185.

Jochle W. 1998. 7th Int. Symp. on Equine Reprod., University of Pretoria, South Africa, July 12 to 17, pp. 1-14.

Joo K, Park W, Lim MJ, Kwon SR and Yoon J. 2011. Serum procalcitonin for differentiating bacterial infection from disease flares in patients with autoimmune diseases. *J Korean Med Sci* 26: 1147-1151.

Jorgensen J H and Ferraro MJ. 2009. Antimicrobial susceptibility testing: A review of general principles and contemporary practices. *Clin Infect Dis* 49:1749-1755.

Jost BH, Songer JG and Billington SJ. 2001. Cloning, expression, and characterization of a neuraminidase gene from Arcanobacterium pyogenes. *Infect Immun* 69:4430-4437.

Jost BH, Songer JG and Billington SJ. 2002. Identification of a second Arcanobacterium pyogenes neuraminidase and involvement of neuraminidase activity in host cell adhesion. *Infect Immun* 70:1106-1112.

Kaczmarowski M, Malinowski E and Markiewicz H. 2004. Influence of various treatment methods on bacteriological findings in cows with puerperal endometritis. *Polish J Vet Sci* 7:171-174.

Kahn CM, Line S and Allen DG. 2005. The Merck Veterinary Manual, Merck; Whitehouse Station, NJ.

Kaker ML, Murray RD and Dobson H. 1984. Plasma hormone changes in cows during induced or spontaneous calvings and the early postpartum period. *Vet Rec* 115:378-382.

Kalmus P, Simojoki H, Orro T, Taponen S, Mustonen K, Holopainen J and Pyorala S. 2014. Efficacy of 5-day parenteral versus intramammary benzylpenicillin for treatment of clinical mastitis caused by gram-positive bacteria susceptible to penicillin *in vitro*. *J Dairy Sci* 97:2155-2164.

Kane JS and Ford AC. 2016. Rifaximin for the treatment of diarrhea-predominant irritable bowel syndrome. *Expert Rev Gastroenterol Hepatol* 10:431-442.

Kaneene JB, Miller R, Herdt TH and Gardiner JC. 1997. The association of serum nonesterified fatty acids and cholesterol, management and feeding practices with peripartum disease in dairy cows. *Prev Vet Med* 31:59-72.

Kaneko K and Kawakami S. 2008. Influence of experimental intrauterine infusion of *Arcanobacterium pyogenes* solution on ovarian activity in cycling cows. *J Vet Med Sci* 70:77-83.

Kaneko K and Kawakami S. 2009. The roles of PGF2α and PGE2 in regression of the corpus luteum after intrauterine infusion of *Arcanobacterium pyogenes* in cows. *Theriogenology* 71:858-863.

Karsch FJ, Battaglia DF, Breen KM, Debus N and Harris TG. 2002. Mechanisms for ovarian cycle disruption by immune/inflammatory stress. *Stress* 5:101-112.

Karstrup CC, Agerholm JS, Jensen TK, Swaro LRV, Klitgaard K, Rasmussen EL, Krogh KM and Pedersen HG. 2017. Presence and localization of bacteria in the bovine endometrium postpartum using fluorescence in situ hybridization. *Theriogenology* 92:167-175.

Kasimanickam R, Duffield T F, Foster RA, Gartley CJ, Leslie KE, Walton J S and Johnson W H. 2005a. The effect of a single administration of cephapirin or cloprostenol on the reproductive performance of dairy cows with subclinical endometritis. *Theriogenology* 63:818-830.

Kasimanickam R, Duffield TF, Foster RA, Gartley CJ, Leslie KE, Walton JS, Johnson WH. 2005b. A comparison of the cytobrush and uterine lavage techniques to evaluate endometrial cytology in clinically normal postpartum dairy cows. *Can Vet J* 46:255-259.

Kasimanickam R, Duffield TF, Foster RA, Gartley CJ, Leslie KE, Walton JS and Johnson WH. 2004. Endometrial cytology and ultrasonography for the detection of subclinical endometritis in postpartum dairy cows. *Theriogenology* 62:9-23.

Kasimanickam RK, Kasimanickam VR, Olsen JR, Jeffress EJ, Moore DA and Kastelic JP. 2013. Associations among serum pro-and anti-inflammatory cytokines, metabolic mediators, body condition, and uterine disease in postpartum dairy cows. *Reprod Biol Endocrinol* 11:103.

Katila T. 1996. Uterine defence mechanisms in the mare. *Anim Reprod Sci* 42: 197-204.

Kaufmann T, Drillich M, Tenhagen B-A and Heuwieser W. 2010a. Correlations between periparturient serum concentrations of non-esterified fatty acids, beta-hydroxybutyric acid, bilirubin, and urea and the occurrence of clinical and subclinical postpartum bovine endometritis. *BMC Vet Res* 6:47

Kaufmann T, Drillich M, Tenhagen B-A, Forderung D and Heuwieser W. 2009. Prevalence of bovine subclinical endometritis 4h after insemination and its effects on first service conception rate. *Theriogenology* 71:385-391.

Kaufmann TB, Westermann S, Drillich M, Plöntzke J and Heuwieser W. 2010b. Systemic antibiotic treatment of clinical endometritis in dairy cows with ceftiofur or two doses of cloprostenol in a 14-d interval. *Anim Reprod Sci* 121:55-62.

Kauppinen K. 1983. Correlation of whole blood concentrations of acetoacetate, B-hydroxybutyrate, glucose and milk yield. *Acta Vet Scand* 24:337-348.

Kayacik V, Salmanoðlu MR, Polat B and Özlüer A. 2005. Evaluation of the corpus luteum size throughout the cycle by ultrasonography and progesterone assay in cows. *Turkish J Vet Anim Sci* 29:1311-1316.

Kebede G, Zenebe T, Disassa H, Tolosa T. 2014. Review on detection of antimicrobial residues in raw bulk milk in dairy farms. *Afr J Basic Appl Sci* 6:87-97.

Kehrli Jr ME, Nonneke BJ and Roth JA. 1989. Alteration in bovine neutrophil function during the periparturient period. *Am J Vet Res* 50:207-214.

Kehrli Jr. ME, Goff JP, Stevens MG and Boone TC. 1991. Effects of granulocyte colony-stimulating factor administration to periparturient cows on neutrophils and bacterial shedding. *J Dairy Sci* 74: 2448-2458.

Kelly JM. 1977. Changes in serum B hydroxybutyrate concentrations in dairy cows kept under commercial farm conditions. *Vet Rec* 101:499-502.

Kelton DF, Leslie KE, Etherington WG, Bonnett BN and Walton JS. 1991. Accuracy of rectal palpation and of a rapid milk progesterone enzymeimmunoassay for determining the presence of a functional corpus luteum in subestrous dairy cows. *Can Vet J* 32:286-291.

Kendall NR, Flint APF and Mann GE. 2009. Incidence and treatment of inadequate postovulatory progesterone concentrations in repeat breeder cows. *Vet J* 181:158-162.

Kennedy PC and Miller RB. 1993. The female genital system. In: Jubb KVF, Kennedy PC, Palmer N, editors. *Pathology of domestic animals*. 4th ed., San Diego: Academic Press; p. 378-87.

Kerns BJ, Morkoc AC, Whitmore HL, McKenna DJ and Gustafsson BK. 1985. Effect of estrogen on neutrophil phagocytosis in ovariectomised cows. In: Proceedings of the 66th Annual Meeting Conference Residential Workers Animal Disease. Abstract, Chicago, p. 11.

Khoshvaghti A, Nazifi S, Vafafar A and Gheisari H R. 2009. Evaluation of serum and milk haptoglobin in some inflammatory diseases of cattle. *Int J Vet Res* 3:31-36.

Killingbeck J and Lamming GE. 1963. Influence of uterine secretions on phagocytosis. *Nature* 198:111-112.

Kim H, Hawley TS, Hawley RG and Baumann H. 1998. Protein tyrosine phosphate 2 moderates signaling by gp130 but is not required for the induction of acute-phase plasma protein genes in hepatic cells. *Mol Cell Biol* 18:1525-1533.

Kim H, Seo JY, Roh KH, Lim JW and Kim KH. 2000. Suppression of NF-KB activation and cytokine production by N-acetylcysteine in pancreatic cell. *Free Radic Biol Med* 29:674-683.

Kim IH, Na KJ and Yang MP. 2005. Immune responses during the peripartum period in dairy cows with postpartum endometritis *J Reprod Dev* 51:757-764

Kim MH, Yang JY, Upadhaya SD, Lee HJ, Yun CH and Ha JK. 2011. The stress of weaning influences serum levels of acute-phase proteins, iron-binding proteins, inflammatory cytokines, cortisol, and leukocyte subsets in Holstein calves. *J Vet Sci* 12:151-157.

Kimura K, Goff JP, Canning P, Wang C and Roth JA. 2014. Effect of recombinant bovine granulocyte colony-stimulating factor covalently bound to polyethylene glycol injection on neutrophil number and function in periparturient dairy cows. *J Dairy Sci* 97:4842-4851.

Kindahl H, Edqvist LE, Larsson K and Malmqvist A. 1982. Influence of prostaglandins on ovarian function postpartum. In: Karg, H., Schallenberger, E. (Eds.), *Factors Influencing Fertility in the Postpartum Cow, Current Topics in Veterinary Medicine and Animal Science*, Vol. 20. Marunus Nijhoff, The Hague, pp. 173-196.

Kindahl H, Fredriksson G, Madej A, Edqvist L-E. 1984. Role of prostaglandins in uterine involution. Proc. 10th International Congress on Animal Reproduction and Artificial Insemination, Vol IV, Urbana-Champaign; 9-16.

Kindahl H, Odensvik K, Aiumlamai S and Fredriksson G. 1992. Utero-ovarian relationships during the bovine postpartum period. *Anim Reprod Sci* 28:363-369.

Klucinski W, Targowski SP, Miernik-Degorska E and Winnicka A. 1990a. The phagocytic activity of polymorphonuclear leukocytes isolated from normal uterus and that with experimentally induced inflammation in cows. *J Vet Med A* 37:506-512.

Klucinski W, Targowski SP, Winnicka A, Miernisk-Degorska E. 1990b. Immunological induction of endometritis model investigations in cow. *Zentralbl Veterinarmed* A 37:148-153.

Knickerboker JJ, ThatcherWW, Foster DB, Wolfenson D, Bartol FF and Caton D. 1984. Uterine production of PGF2α in response to oestradiol-17β in cyclic cattle. *J Anim Sci* 55 (Suppl 1), 364.

Knobloch H, Schroedl W, Turner C, Chambers M and Reinhold P. 2010. Electronic nose responses and acute phase proteins correlate in blood using a bovine model of respiratory infection. *Sensor Actuat B-Chem* 144:81-87.

Knudsen LR, Karstrup CC, Pedersen HG, Angen Ø, Agerholm JS, Rasmussen EL, Jensen TK and Klitgaard K. 2016. An investigation of the microbiota in uterine flush samples and endometrial biopsies from dairy cows during the first 7 weeks postpartum. *Theriogenology* 86:642-650.

Knutti B, Busato A and Ku¨pfer U. 2000. Reproductive efficiency of cows with endometritis after treatment with intrauterine infusions or prostaglandin injections or no treament. *J Vet Med Assoc* 47:609-615.

Kochhar HPS. 1989. Disposition of sulfadimidine and chloramphenicol in blood and uterus of buffalo suffering from dystocia. M.V.Sc. thesis, Punjab Agricultural University, Ludhiana.

Königsson K, Gustafsson H and Kindahl H. 2002. 15-ketodihydro- PGF2α, progesterone and uterine involution in primiparous cows with induced retained placenta and post-partal endometritis treated with oxytetracycline and flunixin. *Reprod Dom Anim* 37:43-51.

Königsson K, Gustafsson H, Gunnarsson A, Kindahl H. 2001. Clinical and bacteriological aspects on the use of oxytetracycline and flunixin in primiparous cows with induced retained placenta and post-partal endometritis. *Reprod Domest Anim* 36:247-256.

Konyves L, Szenci O, Jurkovich V, Tegzes L, Tirián A, Solymosi N, Gyulay G and Brydl E. 2009. Risk assessment and consequences of retained placenta for uterine health, reproduction and milk yield in dairy cows. *Acta Veterinaria Brno* 78:163-172.

Koo HL, Sabounchi S, Huang DB and DuPont HL. 2012. Rifaximin therapy of irritable bowel syndrome. *Clin Med Insights Gastroenterol* 5:31-41.

Korf J, Stoltz A, Verschoor J, De Baetselier P and Grooten J. 2005. The mycobacterium tuberculosis cell wall component mycolic acid elicits pathogen-associated host innate immune response. *Eur J Immunol* 35:890-900.

Kovac G, Tothova C, Nagy O, Seidel H and Konvicna J. 2009. Acute phase proteins and their relation to energy metabolites in dairy cows during the pre- and postpartal period. *Acta Vet Brno* 78:441-447.

Krishnakumar K, Amarnath M, Rajusundaram RC and Chandrahasan C. 2003. Efficacy of white side test for sub-clinical endometritis in crosses bred cows. *Indian J Dairy Sci* 56:119-120.

Krishnan BB, Kumar H, Mehrotra S, Singh SK, Goswamil TK, Khan FA, Patra MK and Islam R. 2015. Effect of leukotriene B4 and oyster glycogen in resolving subclinical endometritis in repeat breeding crossbred cows. *Indian J Anim Res* 49:218-222.

Krishnan BB. 2011. Effect of immunomodulators on the recovery of subclinical endometritis in crossbred cow. M.V.Sc. Thesis. IVRI.

Kristiansen M, Graversen JH, Jacobsen C, Sonne O, Hoffman HJ, Law SK and Moestrup SK. 2001. Identification of the haemoglobin scavenger receptor. *Nature* 409:198-201.

Kru¨ger M, Thi Hien T, Zaremba W and Penka L. 1999. Untersuchungen zum einfluss der proteolytischen enzyme trypsin, chymotrypsin und papain auf euterpathogene mikroorganismen. *Tiera¨rztl Prax* 27:207-215.

Kubar H and Jalakas S. 2002. Pathological changes in the reproductive organs of cows and heifers culled because of infertility. *J Vet Med A Physiol Pathol Clin Med* 49:365-372.

Kuhm AE, Suter D, Felleisen R and Rau J. 2009. Identification of Yersinia enterocolitica at the species and subspecies levels by Fourier transform infrared spectroscopy. *Appl Environ Microbiol* 75:5809-5813.

Kumar A, Gupta HP and Prasad S. 2013. Studies on the immunomodulatory and therapeutic efficacy of neem (Azadirachta indica) on endometritis in repeat breeding crossbred cows. *Indian J Anim Reprod* 34:1-5.

Kumar A. 2013. Evaluation of Immunomodulatory and Therapeutic Efficacy of Turmeric, Neem and Garlic on Endometritisin Repeat Breeding Crossbred Cows. M.V.Sc. Thesis (Veterinary Gynaecology and Obstetrics), G.B. Pant University of Agriculture and Technology, Pantnagar.

Kumar S. 2010. Computation of Judicious Treatment of Endometritisin Cows. Ph.D. Thesis (Animal Reproduction, Gynaecology and Obstetrics), Sher-e-Kashmir University of Agricultural Sciences and Technology, Jammu.

Kümmerer HTK. 2001. Pharmaceuticals in the environment – sources, fate, effects and risks. *Aquat Toxicol* 71:391-392.

Lai YA, Lai IH, Tseng CF, Lee J and Mao SJ. 2007. Evidence of tandem repeat and extra thiol-groups resulted in the polymeric formation of bovine haptoglobin: a unique structure of Hp 2-2 phenotype. *J Biochem Mol Biol* 40:1028-1038.

Lambertz C, Völker D, Janowitz U and Gauly M. 2014. Evaluation of vaginal discharge with the Metricheck device and the relationship to reproductive performance in postpartum dairy cows. *Anim Sci J* 85:848-852.

Lamming GE. 1989. Corpus luteum function in dairy cows and embryo mortality. *J of Reprod Fertil (Supplementary)* 37:245-252.

Lammler G and Durr U. 1967. Treatment of intestinal coccidiosis in rabbits with formosulphathiazole. *Berl Munch Tierarztl Wochenschr* 80:455-458.

Langlois MR and Delanghe JR. 1996. Biological and clinical significance of haptoglobin polymorphism in humans. *Clin Chem* 42:1589-1600.

Larsson B. 1987. Determination of Ovulation by Ultrasound Examination and its Relation to the LH Peak in Heifers. *J Vet Med* 34:749-754.

Laven R, Chambers P and Stafford K. 2012. Using non-steroideal anti-inflammatory drugsaround calving: Maximizing comfort productivity and fertility. *Vet J* 192:8-12.

Laven R. 2019. Uterine infection. NAIDS. 1-4

Lavery K, Gabler C, Day J and Killian G. 2004. Expression of haptoglobin mRNA in the liver and oviduct during the oestrous cycle of cows (*Bos taurus*). *Anim Reprod Sci* 84:13-26.

Lavon Y, Leitner G, Goshen T, Braw-Tal R, Jacoby S and Wolfenson D. 2008. Exposure to endotoxin during estrus alters the timing of ovulation and hormonal concentrations in cows. *Theriogenology* 70:956-967.

Lavon Y, Leitner G, Moallem U, Klipper E, Voet H, Jacoby S, Glick G, Meidan R and Wolfenson D. 2011. Immediate and carryover effects of Gram-negative and Gram-positive toxin-induced mastitis on follicular function in dairy cows. *Theriogenology* 76:942-953.

Lavon Y, Leitner G, Voet H and Wolfenson D. 2010. Naturally occurring mastitis effects on timing of ovulation, steroid and gonadotrophic hormone concentrations, and follicular and luteal growth in cows. *J Dairy Sci* 93:911-921.

Layek SS, Mohanty TK, Kumaresan A, Behera K and Chand S. 2013. Cervical mucus characteristics and periestrual hormone concentration in relation to ovulation time in Zebu (Sahiwal) cattle. *Livest Sci* 152:273-281.

Lazzari G, Duchi R, Colleoni S, Baldazzi L, Benedetti V, Galli A, Luini M, Ferrari M and Galli C. 2011. Le patologie uterine cliniche e subcliniche come causa di infertilitànelle bovine da latte: Studio epidemiologico in due allevamentidellaregioneLombardia. *Large Anim Rev* 17:43-47.

Le Blanc MM. 2012. How to use N-acetylcystine and Tricide® for the treatment of endometritis. In, *Proceedings of the 18th Annual Meeting of the Italian Association of Equine Veterinarians* SIVE, Bologna, Italy. 21-25,

LeBlanc MM and Causey RC, 2009. Clinical and subclinical endometritis in the mare: Both threats to fertility. *Reprod Domest Anim* 44:10-22.

LeBlanc MM. 2010.Advances in the diagnosis and treatment of chronic infectious and post-mating-induced endometritis in the mare. *Reprod Domest Anim* 45(Suppl 2):21-27.

LeBlanc SJ, Duffield TF, Leslie KE, Bateman KG, Keefe GP and Walton WH. 2002a. Defining and diagnosing postpartum clinical endometritis and impact on reproductive performance in dairy cows. *J Dairy Sci* 85:2223-2236.

LeBlanc SJ, Duffield TF, Leslie KE, Bateman KG, Keefe GP, Walton JS and Johnson WH. 2002b. The effect of treatment of clinical endometritis on reproductive performance in dairy cows. *J Dairy Sci* 85:2237-2249.

LeBlanc SJ. 2008. Post partum uterine disease and dairy herd performance. *Vet J* 176:102-114.

LeBlanc SJ. 2014. Reproductive tract inflammatory disease in postpartum dairy cows. *Animal* 8 (Suppl 1):54-63.

Lecchi C, Avallon G, Giurovich MM, Roccabianca P and Ceciliani F. 2009. Extra hepatic expression of the acute phase protein alpha 1-acid glycoprotein in normal bovine tissues. *Vet J* 180:256-258.

Leder L-D. 1970. Diagnostic experiences with the naphthol AS-D chloroacetate esterase reaction. *Blut* 21:1-8.

Lee CS, Gogolin-Ewens K and Brandon MR. 1988. Identification of a unique lymphocyte sub-population in the sheep uterus. *Immunology* 63:159-164.

Lee CS, Meeusen E, Gogolin-Ewens K and Brandon MR. 1992. Quantitative and qualitative changes in the intraepithelial lymphocyte population in the uterus of non-pregnant and pregnant sheep. *Am J Reprod Immunol* 28:90-96.

Lee CS, Wooding FBP and Morgan G. 1997. Quantitative analysis throughout pregnancy of intraepithelial large granular and non-granular lymphocyte distributions in the synepitheliochorial placenta of the cow. *Placenta* 18:675-681.

Lee H. 2013. Procalcitonin as a biomarker of infectious diseases. *Korean J Intern Med* 28:285-291.

Lefebvre RC and Stock AE. 2012. Therapeutic efficiency of antibiotics and prostaglandin F2α in postpartum dairy cows with clinical endometritis: An evidence-based evaluation. *Vet Clin N Am Food Anim Pract* 28:79-96.

Leid JG, Ditto AJ, Knapp A, Shah PN, Wright BD, Blust R, Christensen L, Clemons CB, Wilber JP, Young GW, Kang AG, Panzner MJ, Cannon CL, Yun YH, Youngs WJ, Seckinger NM and Cope EK. 2012. *In vitro* antimicrobial studies of silver carbene complexes: activity of free and nanoparticle carbene formulations against clinical isolates of pathogenic bacteria. *J Antimicrob Chemother* 67:138-148.

Lenz M, Drillich M and Heuwieser W. 2007. Evaluation of the diagnosis of subclinical endometritis in dairy cattle using ultrasound. *Berl Munch Tierarztl Wochenschr* 120:237-244.

Leslie MV and Hansen PJ. 1991. Progesterone-regulated secretion of the serpin-like proteins of the ovine and bovine uterus. *Steroids* 56:589-597.

Leung ST, Cheng Z, Sheldrick EL, Derecka K, Derecka K, Flint AP and Wathes DC. 2001. The effects of lipopolysaccharide and interleukins-1alpha,-2 and -6 on oxytocin receptor expression and prostaglandin production in bovine endometrium. *J Endocrinol* 168:497-508

Leutert C, Kruger SV, Plontzke J and Heuweiser W. 2012. Evaluation of vaginoscopy for the diagnosis of clinical endometritis in dairy cows. *J Dairy Sci* 95:206-212.

Levy A P, Asleh R, Blum S, Levy NS, Miller-Lotan R. and Kalet-Litman S. 2010. Haptoglobin: basic and clinical aspects. *Antioxid Redox Signal* 12:293-304.

Lewis GS. 1997. Health problems of the postpartum cow, uterine health and disorders. *J Dairy Sci* 80:984-994.

Lewis GS. 2003. Steroidal regulation of uterine resis-tance to bacterial infection in livestock. *Reprod Biol End* 1:117-125.

Lewis GS. 2004. Steroidal regulation of uterine immune defences. *Anim Reprod Sci* 82-83:281-294.

Lewis RS. 2001. Calcium signaling mechanisms in T lymphocytes. *Annu Rev Immunol* 19:497-521.

Li J, Zhu W, Liu W, Wu Y and Wu B. 2016. Rifaximin for Irritable Bowel Syndrome: A Meta-Analysis of Randomized Placebo-Controlled Trials. *Medicine (Baltimore)* 95(4):e2534.

Lidbury JA, Cook AK, Steiner JM. 2016. Hepatic encephalopathy in dogs and cats. *J Vet Emerg Crit Care* (San Antonio). 26:471-487.

Lim HJ, Son JK, Yoon HB, Baek KS, Kim TI, Jung YS and Kwon EG. 2014. Physical properties of estrus mucus in relation to conception rates in dairy cattle. *J Embryo Trans* 29:157-161.

Lim SK, Kim H, Lim SK, Ali Ab, Lim YK, Wang Y, Chong SM, Costantini F and Baumman H. 1998. Increased Susceptibility in Hp Knockout Mice During Acute Hemolysis. *Blood* 92:1870-1877.

Lindell JO and Kindahl H. 1983. Exogenous prostaglandin F2 alpha promotes uterine involution in the cow. *Acta Vet Scand* 24:269-274

Lindell JO, Kindahl H, Jansson L and Edqvist LE. 1982. Postpartum release of prostaglandin F2α and uterine involution in the cow. *Theriogenology* 17:237-245.

Lipperheide C, Gothe C, Peterson B and Sommer H. 1997. Nephelometric assay of haptoglobin in blood plasma from cattle, pigs, horses. *Tierarztl* 52:420-426.

Liu IKM and Cheung ATW. 1986. Immunoglobulin and neutrophil defense against uterine infection in mares resistant and susceptible to chronic endometritis: a review. *J Am Vet Med Assoc* 189:700-702.

Liu J, Jiang L, Zhang J, Huang J, Li X and Xu F. 2021. Residue of Rifaximin in milk of lactating dairy cows following administration of intrauterine infusion. *IOP Conf. Ser. Earth Environ Sci* 697 012001.

Liu WJ and Hansen PJ. 1993. Effect of the progesterone-induced serpin-like proteins of the sheep endometrium on natural-killer cell activity in sheep and mice. *Biol Reprod* 49:1008-1014.

Locatelli C, Scaccabarozzi L, Pisoni G, Bronzo V, Casula A, Testa F, Allodi S, Pollera C, Toni F and Moroni P. 2013. *Helcococcus kunzii* and *Helcococcus ovis* isolated in dairy cows with puerperal metritis. *J Gen Appl Microbiol* 59:371-374.

Lock TF. 1982. Distribution of antibiotics in the mare reproductive tract after various routes of administration. *J Reprod Fertil, Supp* 32.

Loncar KD, Ferris RA, McCue PM, Borlee GI, Hennet ML and Borlee BR. 2017. *In vitro* biofilm disruption and bacterial killing using nonantibiotic compounds against gram-negative equine uterine pathogens. *JEVS* 53:94-99.

Lopdell T, Berg MC, Green MP and Back PJ. 2011. Effect of sub-clinical uterine infection on plasma aminoacid concentrations in early lactation dairy cows. *Proc NZ Soc Anim Prod.* 71:291-295.

López-Helguera I, López-Gatius F and García-Ispiero I. 2012. The influence of genital tract status in postpartum period on the subsequent reproductive performance in high producing dairy cows. *Theriogenology* 77:1334-1342.

Loyi T, Kumar H, Nandi S and Patra MK. 2015. Expression of Pathogen Recognition Receptors and Pro-Inflammatory Cytokine Transcripts in Clinical and Sub-Clinical Endometritis Cows. *Anim Biotech* 26:194-200.

Loyi T, Kumar H, Nandi S, Mathapati BS, Patra MK and Pattnaik B. 2013. Differential expression of pro-inflammatory cytokines in endometrial tissue of buffaloes with clinical and sub-clinical endometritis. *Res Vet Sci* 94:336-340.

Lu W, Zabuli J, Kuroiwa T, Tanaka T and Kamomae H. 2011. Effect of intrauterine infusion with liquid paraffin on phagocytes migrating to mucus of external os of the cervix in cows. *Reprod Domest Anim* 46:602-607.

Luyt CE, Guerin V, Combes A, Trouillet JL, Ayed SB, Bernard M, Gibert C and Chastre J. 2005. Procalcitonin kinetics as a prognostic marker of ventilator-associated pneumonia. *Am J Respir Crit Care Med* 171:48-53.

Lyimo ZC, Nielen M, Ouweltjes W, Kruip TA and Van Eerdenburg FJCM. 2000. Relationship among estradiol, cortisol and intensity of estrous behavior in dairy cattle. *Theriogenology* 53:1783-1795.

Mabona U, Viljoen A, Shikanga E, Marston A and Van Vuuren S. 2013. Antimicrobial activity of Southern African medicinal plants with dermatological relevance: From an ethnopharmacological screening approach, to combination studies and the isolation of a bioactive compound. *J Ethnopharmacol* 148:45-55.

Machado VS, Knauer WA, Bicalho MLS, Oikonomou G, Gilbert RO and Bicalho RC. 2012a. A novel diagnostic technique to determine uterine health of Holstein cows at 35 days postpartum. *J Dairy Sci* 95:1349-1357.

Machado VS, Oikonomou G, Bicalho ML, Knauer WA, Gilbert R and Bicalho RC. 2012b. Investigation of postpartum dairy cows' uterine microbial diversity using metagenomic pyrosequencing of the 16S rRNA gene. *Vet Microbiol* 159:460-469.

Madej A, Kindahl H, Larsson K and Edqvist LE. 1986. Sequential hormonal changes in the postpartum dairy cow. *Acta Vet Scand* 27:280-295.

Madej A, Kindahl H, Woyno W, Edqvist LE and Stupnicki R. 1984. Blood levels of 15-keto-13,14-Dihydroprostaglandin F during the postpartum period in primiparous cows. *Theriogenology* 21:279-287.

Madoz LV, Giuliodori MJ, Jaureguiberry M, Plöntzke J, Drillich M and de la Sota RL. 2013. The relationship between endometrial cytology during estrous cycle and cutoff points for the diagnosis of subclinical endometritis in grazing dairy cows. *J Dairy Sci* 96:4333-4339.

Madoz LV, Giuliodori MJ, Migliorisi AL, Jaureguiberry M and de la Sota RL. 2014. Endometrial cytology, biopsy, and bacteriology for the diagnosis of subclinical endometritis in grazing dairy cows. *J Dairy Sci* 97:195-201.

Madoz LV, Sota RL De La, Suzuki K, Heuwieser W and Drillich M. 2010. Use of hysteroscopy for the diagnosis of postpartum clinical endometritis in dairy cows. *Vet Rec* 167:142-143.

Madsen M, Moller HJ, Nielsen MJ, Jacobsen C, Graversen JH, van den Berg T and Moestrup SK. 2004. Molecular characterization of the haptoglobin.hemoglobin receptor CD163. Ligand binding properties of the scavenger receptor cysteine-rich domain region. *J Biol Chem* 279:51561-51567.

Magness RR, Huie JM, Hoyer GL, Huecksteadt TP, Reynolds LP, Seperich GJ, Whysong G and Weems CW. 1981. Effect of chronic ipsilateral or contralateral intrauterine infusion of prostaglandin E2 (PGE2) on luteal function of unilaterally ovariectomized ewes. *Prostaglandins Med* 6:389-401.

Magni E, Ferrari M, Hickel R, Huth KC and Ilie N. 2008. Effect of ozone gas application on the mechanical properties of dental adhesives bonded to dentin. *Dent Mater* 24:1428-1434.

Mahmood T, Abbas M, Ilyas S, Afzal N, Nawaz R. 2016. Quantification of fluoroquinolone (enrofloxacin, norfloxacin and ciprofloxacin) resi-dues in cow milk. *Int J Chem Biochem Sci* 10:10-15.

Makki M, Ahmadi M R, Gheisari H R and Nazif, S. 2017. Cure rate of postpartum endometritis after different treatments in high produce dairy cows. *Comp Clin Path* 26:1-8.

Makki M, GheisarI HR and Ahmadi MR. 2016. Effect of different intrauterine oxytetracycline treatment on reproductive performance of dairy cows with clinical endometritis and determination of oxytetracycline residues in milk. *Acta Vet Eurasia* 42:80-88.

Malinowski E, Lassa H, Markiewicz H, Kaptur M, Nadolny M, Niewitecki W and Zietara J. 2011. Sensitivity to antibiotics of *Arcanobacterium pyogenes* and *Escherichia coli* from the uteri of cows with metritis/endometritis. *Vet J* 187:234-238.

Mandhwani R, Bhardwaz A, Kumar S, Shivhare M and Aich R. 2017. Insights into bovine endometritis with special reference to phytotherapy, *Vet World* 10:1529-1532.

Manjhi VK, Bajaj NK, Shukla SP, Shrivastava N, Kaurav PS and Thakur RK. 2019. Bacterial isolation and antibiogram of uterine lavage from repeat breeder cows suffering from subclinical endometritis. *J Anim Res* 9:293-296.

Mann GE and Lamming GE. 1995. Progesterone inhibition of the development of the luteolytic signal in cows. *Reproduction* 104:1-5.

Mann S, Siler JD, Jordan D and Warnick LD. 2011. Antimicrobial susceptibility of fecal Escherichia coli isolates in dairy cows following systemic treatment with ceftiofur or penicillin. *Foodborne Pathog Dis* 8:861-867.

Mansour MM, Hendawy AO and Zeitoun MM. 2016. Effect of mastitis on luteal function and pregnancy rates in buffaloes. *Theriogenology* 86:1189-1194.

Mari G, Lacono E, Toni F, Predieri PG and Merlo B. 2012. Evaluation of the effectiveness of intrauterine treatment with formosulphathiazole of clinical endometritis in postpartum dairy cows. *Theriogenology* 78:189-200.

Marino B, Quintela LA, Becerra JJ, Barrio L, Mocinos JE and Guillin JL. 2017. Agreement between postmortem endometrial cytology, biopsy and bacteriology in culled dairy cows. *Anim Reprod* 14:1024-1033.

Markusfeld O. 1984. Factors responsible for post parturient metritis in dairy cattle. *Vet Rec* 114:539-542.

Markusfeld O. 1987. Periparturient traits in seven high dairy herds. Incidence rates, association with parity, and interrelationships among traits. *J Dairy Sci* 70:158-166.

Martínez-Cortés I, Rosiles R, Gutierrez L, Díaz D, Sumano H. 2014. Amoxicillin resides in milk of holstein cows with double or triple intramammary administration of amoxicillin-potassium clavula-nate. *Int J Appl Res Vet Med* 12:186-192

Marusi A, Allegri M, Mararsi G, Orsi G and Ubaldi A. 1999: I lipoperossidi nella profilassi e terapia della metrite e nel miglioramento della fertilita‘ nella bovina da latte, Atti Soc. It. Buiatria, Bologna.

Marusi A, Ubaldi A, Fusari A, Marasi G and Isnenghi F. 2000. Haptoglobin response in dairy cow metritis treatment with lipohydroperoxides. *XXI World Buiatrics Congress*, Punta del Este Uruguay, December 2000. pp. 4-8.

Masera J, Gustafsoon BK, Affiefy MM, Stowe CM and Bergt GP. 1980b. Disposition of oxytetracycline in the bovine genital tract. Systemic versus intrauterine administration. *J Am Vet Med Assoc* 178:1099-1102.

Masera J, Gustafsson BK and Afiefy MM. 1980a. Blood plasma and uterine tissue concentrations of sodium penicillin G in cows at intramuscular vs. intrauterine administration. *Proceeding 9th ICAR*. Madrid. 42: 687-690.

Mateus L, Lopes da Costa L, Bernardo F and Silva J.R. 2002a. Influence of puerperal uterine infection on uterine involution and postpartum ovarian activity in dairy cows. *Reprod Dom Anim* 37:31-35.

Mateus L, Lopes da Costa L, Carval H, Serra P and Robolo Silva J. 2002b. Blood and intrauterine leukocyte profile and function in dairy cows that spontaneously recovered from postpartum endometritis. *Reprod Dom Anim* 37:176-181.

Mather EC. 1975. In vivo uterine lumen pH values of the bovine. *Theriogenology* 3:113-119.

Mattmiller SA, Carlson BA and Sordillo LM. 2013. Regulation of inflammation by selenium and selenoproteins: impact on eicosanoid biosynthesis. *J Nutr Sci* 2e28.

Max A, Jurka P, Witkowski M, Borryczko Z and Bostedt H. 1997. Kritischer Vergleich zwischen klisch und ultrasonographisch erfassten ovarbefunden im Interoestrum des Rindes. *Tierarztl Prax* 25:207-211.

May TW. 1996. Investigations relating to intrauterine pH and postpartum endometritis in dairy cattle, with particular reference to Actinomyces pyogenes. DBR dissertation, University of Liverpool, Liverpool.

Mazzaferro EM, Rudloff E and Kirby R. 2002. The role of albumin replacement in the critically ill veterinary patient. *JVECC* 12:113-124.

McClary D, Putnam M, Wright J and Sartin, J. 1989. Effect of early postpartum treatment with prostaglandin F2α on subsequent fertility in the dairy cow. *Theriogenology* 31:565-570.

McCoy MA, Lennox SD, Mayne CS, McCaughey WJ, Edgar HWJ, Catney DC, Verner M, Mackey DR and Gordon AW. 2006. Milk progesterone profiles and their relationship with fertility, production and disease in dairy cows in Northern Ireland. *Anim Sci* 82:213-222.

McCracken JA, Custer EE and Lamsa JC. 1999. Luteolysis: a neuroendocrine-mediated event. *Physiological Rev* 79:263-323.

McDougall S, Hussein H, Aberdein D, Buckle K, Roche J, Burke C, Mitchell M and Meier S. 2011. Relationships between cytology, bacteriology and vaginal discharge scores and reproductive performance in dairy cattle. *Theriogenology* 76:229-240.

McDougall S, McCaulay R and Compton C. 2007. Association between endometritis diagnosis using a novel intravaginal device and reproductive performance in dairy cattle. *Anim Reprod Sci* 99:9-23.

McDougall S. 2001. Effect of intrauterine antibiotic treatment on reproductive performance of dairy cows following periparturient disease. *NZVJ* 49:150-158.

McLaughlin C L, Stanisiewski EP, Risco CA, Santos JE, Dahl GE, Chebel RC, LaGrow C, Daugherty C, Bryson, D.Weigel L, Hallberg J and Lucas MJ. 2013. Evaluation of ceftiofur crystalline free acid sterile suspension for control of metritis in high-risk lactating dairy cows. *Theriogenology* 79:725-734.

Medina-Bastidas D, Camacho-Arroyo I and García-Gómez E. 2022. Current findings in endometrial microbiome: impact on uterine diseases. *Reproduction* 163:R81-R96.

Meira Jr. EBS, Henriques LCS, Sa LRM and Gregory L. 2012. Comparison of ultrasonography and histopathology for the diagnosis of endometritis in Holstein-Friesian cows. *J Dairy Sci* 95: 6969-6973.

Meisner M, Lohs T, Huettemann E, Schmidt J, Hueller M, Reinhart K. 2001. The plasma elimination rate and urinary secretion of procalcitonin in patients with normal and impaired renal function. *Eur J Anaesthesiol* 18:79-87.

Meisner M. 2002 Pathobiochemistry and clinical use of procalcitonin. *Clin Chim Acta* 323:17-29.

Melcher Y, Prunner I and Drillich M. 2014. Degree of variation and reproducibility of different methods for the diagnosis of subclinical endometritis. *Theriogenology* 82:57-63.

Melkus E, Witte T, Walter I, Heuwieser W and Aurich C. 2012. Investigations on the endometrial response to intrauterine administration of N-acetylcysteine in oestrus mares. *Reprod Dom Anim*, doi: 10.1111/rda.12131, 1-7.

Methai A, Rajasundaram RC, Veerapandian C and Ahmad M. 2005. Intrauterine plasma treatment for endometritis in Jersey crossbred cows. *Indian J Anim Reprod* 26:7-10.

Methai A. 1999. Efficacy of lipopolysaccharide and plasma in the treatment of endometritis in cows. M.V. Sc. Thesis, Madras Veterinary College.

Metris A, George SM and Baranyi J. 2006. Use of optical density detection times to assess the effect of acetic acid on single-cell kinetics. *Appl Environ Microbiol* 72:6674-6679.

Mikeska JC and Williams GL. 1988. Timing of preovulatory endocrine events, estrus and ovulation in Brahman× Hereford females synchronized with norgestomet and estradiol valerate. *J Anim Sci* 66:939-946.

Mileva R, Karadaev M, Fasulkov I, Petkova T, Rusenova N, Vasilev N and Milanova A. 2020. Oxytetracycline pharmacokinetics after intramuscular administration in cows with clinical metritis associated with *Trueperella pyogenes* infection. *Antibiotics* 9:392.

Mileva R, Karadaev M, Fasulkov I, Rusenova N, Vasilev N and Milanova A. 2022. Oxytetracycline persistence in uterine secretion after intrauterine administration in cows with metritis. *Animals (Basel)* 28:12:1922.

Miller AN, Williams EJ, Sibley K, Herath S, Lane EA, Fishwick J, Nash DM, Rycroft AN, Dobson H, Bryant CE and Sheldon IM. 2007. The effects of *Arcanobacterium pyogenes* on endometrial function *in vitro*, and on uterine and ovarian function in vivo. *Theriogenology* 68:972-980.

Miller GE and Rouse G. 1970. Passage of drugs from the bovine uterus into blood, milk and urine. *J Dairy Sci* 53:653-657.

Miller HV, Kimsey PB and Kendrick JW. 1980. Endometritis of dairy cattle: diagnosis, treatment, and fertility. *Bov Pract* 15:13-23.

Miller RB. 1977. A summary of some of the pathogenetic mechanisms involved in bovine abortion. *Can Vet J* 18:87.

Milosh P, Ivan P, Slobodanka V, Dobrila J., Voislav P, Mile D, Marina R, Svetlana N and Milan M. 2013. Comparison of microbiological and endoscopy diagnostic of uterine disease in infertility of dairy cattle. *Proceedings 4th International Scientific Meeting, 6-8 September, Republic of Macedonia*, pp 51-52.

Mitchell G, Liu IKM, Perryman LE, Stabenfeldt GH and Hughes JPH. 1982. *J Reprod Fertil* (Suppl.) 32:161-166.

Moffatt RJ, Bazer FW, Hansen PJ, Chun PW and Roberts RM. 1987. Purification, secretion and immunocytochemical localisation of uterine milk proteins, major progesterone-induced proteins in uterine secretions of the sheep. *Biol Reprod* 36:419-430.

Mohammed ZA, Mann GE and Robinson RS. 2019. Impact of endometritis on post-partum ovarian cyclicity in dairy cows. *Vet J* 248:8-13.

Mollett TA, Elmore RG, Blanchard TL and Berg JN. 1985. Effects of intrauterine infusion of Escherichia coli endotoxin in anestrous and steroid treated pony mares. *Theriogenology* 23:597-606.

Momont HW and Seguin BE. 1985. Prostaglandin therapy and the postpartum cow. *Bov Pract* 17:89-92.

Montgomery ME. 1988. Reproductive performance of dairy cows following treatment with a single or two sequential doses of cloprostenol 26 and/or 40 days postpartum. *Theriogenology* 29:565-575.

Mordak R. 2008. Usefulness of haptoglobin for monitoring the efficiency of therapy of fetal membrane retention in cows. *Med Wet* 64:434-443.

Morimatsu M, Syuto B, Shimada N, Fujinaga T, Yamamoto S, Saito M and Naiki M. 1991. Isolation and characterization of bovine haptoglobin from acute phase sera. *J Biol Chem* 266:11833-11837.

Morrow DA. 1969. Postpartum ovarian activity and involution of the uterus and cervix in dairy cattle. *Vet Scope XIV*, 2-13.

Mortimer RC, Farin PW and Stevens RD. 1997. Reproductive examination of the non-pregnant cow. In: Youngquist, R.S. (Ed.), *Current Therapy in Large Animal Theriogenology*. W.B. Saunders, Philadelphia, PA, pp. 268-275.

Moyes RB, Reynolds J and Breakwell DP. 2009. Differential Staining of Bacteria: Gram Stain. *Curr Protoc Microbiol* 15: p 3.

MSD Manual Veterinary Manual. Utrine diseases in dairy cattle. Web site. https://www.msdvetmanual.com/reproductive-system/metritis-in-large-animals/acute-puerperal-metritis-in-large-animals. Accessed September 12, 2023.

Mufti MMR, Alam MK, Sarker M.S, Bostami ABMR and Das NG. 2010. Study on factors affecting the conception rate in Red Chittagong cows. *Bangladesh J Anim Sci* 39:52-57.

Murugavel K and López-Gatius F. 2009. Newtonian behaviour of the vaginal fluid as a risk indicator of reduced fertility in cows. *Indian Vet J* 86:1288-1289.

Mylonaki E, Voutsas V, Antoniou D, Papakosta D, Kontakiotis T, Skordalaki A and Christaki P. 2008. Hydrocarbon pneumonitis following liquid paraffin aspiration during a fire eating performance: A case report. *J Med Case Reports* 2:214.

Nagant C, Tré-Hardy M, El-Ouaaliti M, Savage P, Devleeschouwer M, Dehaye JP. 2010. Interaction between tobramycin and CSA-13 on clinical isolates of Pseudomonas aeruginosa in a model of young and mature biofilms. *Appl Microbiol Biotechnol* 88:251-263.

Naidu Krishna S, Govender P and Adam JK. 2015. Nano silver particles in biomedical and clinical applications. *J Pure Appl Microbiol* 9:103-112.

Nak Y, Dagalp SB, Cetin C, Nak D, Alkan F, Borum E and Tuna B. 2011. Course and severity of postpartum metritis cases following antibiotic and PGF2α administration in postpartum metritis cows infected with BoHV-4. *Transbound Emerg Dis* 58:31-36.

Nakagawa H, Yamamoto O, Oikawa S, Higuchi H, Watanabe A and Katoh N. 1997. Detection of serum haptoglobin by enzyme-linked immunosorbent assay in cows with fatty liver. *Res Vet Sci* 62:137-141.

Nakajima Y, Mikami O, Yoshioka M, Motoi Y, Ito T, Ishikawa Y, Fuse M, Nakano K and Yasukawa K. 1997. Elevated levels of tumor necrosis factor-α,(TNF-α) and interleukin-6 (IL-6) activities in the sera and milk of cows with naturally occurring coliform mastitis. *Res Vet Sci* 62:297-298.

Nakamura TJ, Moriya T, Inoue S, Shimazoe T, Watanabe S, Ebihara S and Shinohara K. 2005. Estrogen differentially regulates expression of Per1 and Per2 genes between central and peripheral clocks and between reproductive and nonreproductive tissues in female rats. *J Neuroscience Res* 82:622-630.

Nakao T, Gamal A, Osawa T, Nakada K, Moriyoshi M and Kawata K. 1997. Postpartum plasma PGF metabolite profile in cows with dystocia and/or retained placenta, and effect of fenprostalene on uterine involution and reproductive performance. *J Vet Med Sci* 59:791-794.

Nakao T, Moriyoshi M and Kawata K. 1988. Effect of postpartum intrauterine treatment with 2% polyvinyl-pyrrolidone-iodine solution on reproductive efficiency in cows. *Theriogenology* 30:1033-1043.

Narducci D. 2007. An introduction to nanotechnologies: What's in it for us? *Vet Res Commun* 31:131-137.

Naumann D, Helm D and Labischinski H. 1991. Microbiological characterizations by FT-IR spectroscopy. *Nature* 351:81-82.

Nazhat SA, Kitahara G, Kozuka N, Mido S, Sadawy M, Ali HE and Osawa T. 2018. Associations of periparturient plasma biochemical parameters, endometrial leukocyte esterase and myeloperoxidase, and bacterial detection with clinical and subclinical endometritis in postpartum dairy cows. *J Vet Med Sci* 80:302-310.

Nazifi S, Khoshvaghti A and Gheisari HR. 2008. Evaluation of serum and milk amyloid A in some inflammatory diseases of cattle. *Iran J Vet Res* 9:222-226.

Nazzaro F, Fratianni F, de Martino L, Coppola R and de Feo V. 2013. Effect of essential oils on pathogenic bacteria. *Pharmaceuticals* 6:1451-1474.

Negi V. 2022. Studies on ovarian and endocrine profile in endometritis affected cycling cows. MVSc thesis submitted to Himachal Pradesh Agricultural University. Palampur.

Nehru DA, Dhaliwal GS, Jan MH, Cheema RS and Kumar S. 2019. Clinical efficacy of intrauterine cephapirin benzathine administration on clearance of uterine bacteria and subclinical endometritis in postpartum buffaloes. *Reprod Dom Anim* 54:317-324.

Ng QX, Loke W, Foo NX, Mo Y, Yeo WS and Soh AYS. 2019. A systematic review of the use of rifaximin for *Clostridium difficile* infections. *Anaerobe* 55:35-39.

Nicholas R, Ayling R and McAuliffe L. 2008. Reproductive diseases of cattle. In *Mycoplasma Diseases of Ruminants* p 208-224.

Niedweski I and Threlfall WR. 1980. The influence of days postpartum on chloramphenicol absorption from bovine uterus. *Theriogenology* 14:319-321.

Nielen M, Aarts MGA, Ad GM, Jonkers T, Wensing YH and Schukken YH. 1994. Evaluation of two cowside tests for the detection of subclinical ketosis in dairy cows. *Can Vet J* 35:229-232.

Nielsen BH, Jacobsen S, Anderson PH, Niewold TA and Heegaard PMH. 2004. Acute phase protein concentrations in serum and milk from healthy cows, cows with clinical mastitis and cows with extramammary inflammatory conditions. *Vet Rec* 154:361-365.

Nielsen JM. 2005. Endometritis in the mare: A diagnostic study comparing cultures from swab and biopsy. *Theriogenology* 64:510-518

Nightingale CR, Sellers MD AND Ballou MA. 2015. Elevated plasma haptoglobin concentrations following parturition are associated with elevated leukocyte responses and decreased subsequent reproductive efficiency in multiparous Holstein dairy cows. *Vet Immunol Immunopathol* 164:16-23.

Nijsten MW, Olinga P, The TH, de Vries EG, Koops HS, Groothuis GM, Limburg PC, ten Duis HJ, Moshage H, Hoekstra HJ, Bijzet J and Zwaveling JH. 2000. Procalcitonin behaves as a fast responding acute phase protein in vivo and *in vitro. Crit Care Med* 28:458-461.

Nino-Soto MI, Heriazon A, Quinton M, Miglior F, Thompson K and Mallard BA. 2008. Differential gene expression of high and low immune responder Canadian Holstein dairy cows. *Dev Biol* 132:315-320.

Nisha AR. 2008. Antibiotic residues—a global health hazard. *Vet World* 1:375-377.

Noakes DE, Parkinson TJ and England GCW. 2019. Veterinary Reproduction and Obstetrics (10th edition). Elsevier Ltd.

Noakes DE, Parkinson TJ, England GCW and Arthur GH. 2002. *Arthur's Veterinary Reproduction and Obstetrics*, eighth ed. Elsevier Sci. Ltd, pp. 399-408.

Noakes DE, Wallace L and Smith GR. 1991. Bacterial flora of the uterus of cows after calving on two hygienically contrasting farms. *Vet Rec* 128:440-442.

Oberhoffer M, Vogelsang H, Jager L and Reinhart K. 1999. Katacalcin and calcitonin immunoreactivity in different types of leukocytes indicate intracellular procalcitonin content. *J Crit Care* 14:29-33.

Oberreuter H, Seiler H and Scherer S. 2002. Identification of coryneform bacteria and related taxa by Fourier-transform infrared (FT-IR) spectroscopy. *Int J Syst Evol Microbiol* 52:91-100.

Ocal H, Yuksel M and Ayar A. 2004. Effects of gentamicin sulfate on the contractility of myometrium isolated from non-pregnant cows. *Anim Reprod Sci* 84:269-277.

Odensvik K, Aiumlamai S, Fredriksson G and Kindahl H. 1992. Pharmacokinetics of flunixin meglumine and its effect during the postpartum period in cattle. In: *Proceedings of the 12th ICAR*, The Hague, pp. 84-86.

Odensvik K, Fredriksson G, 1993: The effect of intensive flunixin treatment during the post partum period in the bovine. *J Vet Med A* 40:561-568.

Ohtani S, Okuda K, Nishimura K and Mohri S. 1993. Histological changes in bovine endometrium during the estrous cycle. *Theriogenology* 39:1033-1042.

Okano A and Tomizuka T. 1987. Ultrasonic observations of postpartum uterine involution in the cow. *Theriogenology* 27:369-376.

Olivos-Garcia A, Carrero JC, Ramos E, Nequiz M, Tello E, Monfort I and Perez-Tamayo R. 2007. Late experimental amebic liver abcess in hamster is inhibited by cyclosporine and acetylcysteine. *Exp Mol Pathol* 82:310-315.

Olson JD, Ball L, Mortimer RG, Farin PW, Adney WS and Huffman EM. 1984. Aspects of bacteriology and endocrinology of cows with pyometra and retained fetal membranes. *American J Vet Res* 45:2251-2255.

Olson JD, Bretzlaff KN, Mortimer RG and Ball, L., 1986. The metritis-pyometra complex. In: Morrow, D.A. (Ed.), *Current Therapy in Theriogenology*. W.B. Saunders Co., Philadelphia, PA, pp. 227-236.

Opsomer G, Gröhn YT, Hertl J, Coryn M, Deluyker H and de Kruif A. 2000. Risk factors for postpartum ovarian dysfunction in high producing dairy cows in Belgium: a field study. *Theriogenology* 53:841-857.

Oral H, Sozman M, Serin G and Kaya S. 2009. Comparison of the cytobrush technique, vaginoscopy, and transrectal ultrasonography methods for the diagnosis of postpartum endometritis in cows. *J Anim Vet Adv* 8:1252-1255.

Osborne TM. 2003. An evaluation of metabolic function in transition dairy cows supplemented with Rumensin premix or administered a rumensin controlled-release capsule. MSc dissertation, University of Guelph.

Otero C, De Ruiz CS, Ibañez R, Wilde OR, de Ruiz Holgado AAP and Nader-Macias ME. 1999. Lactobacilli and enterococci isolated from the bovine vagina during the estrous cycle. *Anaerobe* 3:305-307.

Ott RS, Bretzlaff KN and Hixon JE. 1986. Comparison of palpable corpora lutea with serum progesterone concentrations in cows. *J Am Vet Med Assoc* 188:1417-1419.

Overbeck W, Jäger K, Schoon H-A and Witte T. 2013. Comparison of cytological and histological examinations in different locations of the equine uterusdan in vitro study. *Theriogenology* 79:1262-1268.

Overbeck W, Witte T and Heuwieser W. 2011. Comparison of three diagnostic methods to identify subclinical endometritis in mares. *Theriogenology* 75:1311-1318.

Overton M and Fetrow J. 2008. Economics of postpartum uterine health. Pages 39–44 in *Proc. Dairy Cattle Reproduction Council Conv., Omaha*, NE. Dairy Cattle Reproduction Council, Champaign, IL.

Owis M, Sharad KS, Shehbaz A and Saleemuddin M. 2005. Antibacterial efficacy of Withania somnifera ashwa-gandha an indigenous medicinal plant against experimental murine salmonellosis. *Phytomedicine* 12:229-235.

Ozawa T, Kiku Y, Mizuno M, Inumaru S, Kushibiki S, Shingu H, Matsubara T, Takahashi H and Hayashi T. 2012. Effect of intramammary infusion of rbGM-CSF on SCC and expression of polymorphonuclear neutrophil adhesion molecules in subclinical mastitis cows. *Vet Res Commun* 36:21-27.

Painuly B. 2019. Isolation and identification of predominant bacteria from genital tract of repeat breeder crossbred cows suffering from subclinical endometritis. MVSc Thesis. Department of Veterinary Gynaecology and Obstetrics, G.B. Pant University of Agriculture and Technology, Pantnagar - 263145 (Uttarakhand)

Paisley LG, Mickelsen WOD and Anderson PB. 1986. Mechanisms and therapy for retained fetal membranes and uterine infections of cows: a review. *Theriogenology* 25:353-381.

Pak H, Yokota S, Teplitz R, Shaw S and Werner J. 1980. Rapid staining techniques employed in fine needle aspirations of the lung. *Acta Cytol* 25:178-184.

Palenik T, Dolezel R, Kratochvil J, Cech S, Zajic J, Jan Z and Vyskocil M. 2009. Evaluation of rectal temperature in diagnosis of puerperal metritis in dairy cows . *Vet Med Czech* 54:149-155.

Palmer CW. 2008. Postpartum endometritis: *Current concepts in diagnosis and treatment*. 29th World Vet. Cong. 241-250.

Panangala VS, Fish NA and Barnum DA. 1978. Microflora of the cervico-vaginal mucus of repeat breeder cows. *Can Vet J* 19:83.

Paradelis AG, Triantaphyllidis C, Tarlatzis BC, Tsouras JS, El-Messidi MM and Papaloucas A.C. 1982. Interaction of aminoglycoside antibiotics with oxytocin and prostaglandin E2 on uterine contractility. *Methods Find Exp Clin Pharmacol* 4:343-345.

Park IH, Lee SH, Yu ST and Oh YK. 2014. Serum procalcitonin as a diagnostic marker of neonatal sepsis. *Korean J Pediatr* 57:451-456.

Parrilla-Hernandez S, Ponthier J, Franck TY, Serteyn DD and Deleuze SC. 2014. High concentrations of myeloperoxidase in the equine uterus as an indicator of endometritis. *Theriogenology* 81:936-940.

Pascottini OB, Dini P, Hostens M, Ducatelle R and Opsomer G. 2015. A novel cytologic sampling technique to diagnose subclinical endometritis and comparison of staining methods for endometrial cytology samples in dairy cows. *Theriogenology* 84:1438-1446.

Pascottini OB, Hostens M, Sys P, Vercauteren P and Opsomer G. 2017. Cytological endometritis at artificial insemination in dairy cows: Prevalence and effect on pregnancy outcome. *J Dairy Sci* 100:588-597.

Pascottini OB, Spricigo JFW, Van Schyndel SJ, Mion B, Rousseau J, Weese JS and LeBlanc SJ. 2021. Effects of parity, blood progesterone, and non-steroidal anti-inflammatory treatment on the dynamics of the uterine microbiota of healthy postpartum dairy cows. *PloS One* 16: p 0233943.

Pascottini OB, Van Schyndel SJ, Spricigo JFW, Rousseau J, Weese JS and LeBlanc SJ. 2020. Dynamics of uterine microbiota in postpartum dairy cows with clinical or subclinical endometritis. *Sci Rep* 10, 12353.

Pascottini OB. 2016. Subclinical Endometritis in Dairy Cattle: A Practical Approach. [PhD Thesis]. Department of Reproduction, Obstetrics and Herd Health, Faculty of Veterinary Medicine, Ghent University.

Paufler S. 1969. Intrauterine application of Bykocillin-M (Neomycin plus penicillin) suspension at the time of insemination. *Tierarztt Umsch* 24:588 and 590-591.

Pavlou AK, Magan N, McNulty C, Jones J, Sharp D, Brown J and Turner AP. 2002. Use of an electronic nose system for diagnoses of urinary tract infections. *Biosens Bioelectron* 17:893-899.

Pedersoli WM, Fazeli MH, Haddad NS, Ravis WR and Carson RL. 1985. Endometrial and serum gentamicin concentrations in pony mares given repeated intrauterine infusions. *Am J Vet Res* 46:1025-1028.

Peltier MR, Liu WJ and Hansen PJ. 2000. Regulation of lymphocyte proliferation by uterine serpin: interlukin-2 mRNA production, CD25 expression and responsiveness to interlukin. *Proc Soc Exp Biol* 223:75-81.

Peng Y, Wang Y, Hang S and Zhu W. 2013. Microbial diversity in uterus of healthy and metritic postpartum Holstein dairy cows. *Folia Microbiol (Praha).* 58:593-600.

Peralta S, Cottone C, Doveri T, Almasio PL and Craxi A. 2009. Small intestine bacterial overgrowth and irritable bowel syndrome-related symptoms: experience with Rifaximin. *World J Gastroenterol.* 15:2628-2631.

Perry GA and Perry BL. 2008. Effect of preovulatory concentrations of estradiol and initiation of standing estrus on uterine pH in beef cows. *Dom Anim Endocrinol* 34:333-338.

Perry GA, Smith MF, Lucy MC, Green JA, Parks TE, MacNeil MD, Roberts AJ and Geary TW. 2005. Relationship between follicle size at insemination and pregnancy success. In: *Proceedings of the National Academy of Sciences* 102:5268-5273.

Perry GA, Smith MF, Roberts AJ, MacNeil, MD and Geary TW. 2007. Relationship between size of the ovulatory follicle and pregnancy success in beef heifers. *J Anil Sci* 85:684-689.

Peter AT, Bosu WTK and deDecker RJ. 1989. Suppression of preovulatory luteinising hormone surges in heifers after intrauterine infusions of *Escherichia coli* endotoxin. *Am J Vet Res* 50:368-373.

Peter AT, Bosu WTK and Gilbert R.O. 1990. Absorption of Escherichia coli endotoxin (lipopolysaccharide) from the uteri of postpartum dairy cows. *Theriogenology* 33:1011-1014.

Peter AT, Bosu WTK and Luker CW. 1987. Plasma endotoxin and concentrations of stable metabolites of prostacyclin, thromboxane A2, and prostaglandin E2 in postpartum cows. *Prostaglandin* 34:15-28.

Peter AT, Jarratt GM and Hanlon DW. 2011. Accuracy of diagnosis of clinical endometritis with Metricheck in postpartum dairy cows. *Clinical Theriogenology* 3:461-465.

Petersen H H, Nielsen J P and Heegaard P M. 2004. Application of acute phase protein measurements in veterinary clinical chemistry. *Vet Res* 35:163-187.

Philippidis P, Mason JC, Evans BJ, Nadra I, Taylor KM, Haskard DO and Landis RC. 2004. Hemoglobin scavenger receptor CD163 mediates interleukin-10 release and heme oxygenase-1 synthesis: antiinflammatory monocyte-macrophage responses in vitro, in resolving skin blisters *in vivo*, and after cardiopulmonary bypass surgery. *Circ Res* 94:119-126.

Pierce-Hendry SA and Dennis J. 2010. Bacterial culture and antibiotic susceptibility testing. *Compend Contin Educ Vet* 32:E1–5; quiz E6.

Pietrocola G, Valtulina V, Rindi S, Jost BH and Speziale P. 2007. Functional and structural properties of CbpA, a collagen-binding protein from Arcanobacterium pyogenes. *Microbiology* 153:3380-3389.

Pleticha S, Drillich M and Heuwieser W. 2009. Evaluation of the Metricheck device and the gloved hand for the diagnosis of clinical endometritis in dairy cows. *J Dairy Sci* 92:5429-5435.

Plontzke J, Madoz L, De la Sota R, Drillich M and Heuwieser W. 2010. Subclinical endometritis and its impact on reproductive performance in grazing dairy cattle in Argentina. *Anim Reprod Sci* 122:52-57.

Plontzke J, Madoz L, De la Sota R, Heuwieser W and Drillich M. 2010. Prevalence of clinical endometritis and its impact on reproductive performance in grazing dairy cattle in Argentina. *Reprod Dom Anim* 46:520-526.

Pohl A, Bertulat S, Borchardt S, Burfeind O and Heuwieser W. 2016. Randomized, controlled clinical trial on the efficacy of nonsteroidal antiinflammatory drugs for the treatment of acute puerperal metritis in dairy cows. *J Dairy Sci* 99:8241-8249.

Pohler KG, Geary TW, Atkins JA, Perry GA, Jinks EM and Smith MF. 2012. Follicular determinants of pregnancy establishment and maintenance. *Cell Tissue Res* 349:649-664.

Polat B, Cengiz M, Colak A and Cannazik O. 2015. Comparison of Intrauterine Ozone and Rifaximine Treatment in Cows with Subclinical Endometritis, *Kafkas Universitesi Veteriner Fakultesi Dergisi*, 21:773-776.

Polyantsev NI and Popov Yu N. 1972. Diagnosis and treatment of occult endometritis in cows. *Vastnek Selskoknozyanstvennoi Nauki* 1:22-26.

Ponziani FR, Pecere S, Lopetuso L, Scaldaferri F, Cammarota G and Gasbarrini A. 2016. Rifaximin for the treatment of irritable bowel syndrome - a drug safety evaluation. *Expert Opin Drug Saf* 15:983-991.

Popov Yu N. 1969. Diagnosis *of occult endometritis in cows (Using white side test on cervical mucus) Vet Moscow* 4:85-87.

Pothmann H, Prunner I, Wagener K, Jaureguiberry M, de la Sota RL, Erber R, Aurich C, Ehling-Schulz M and Drillich M. 2015. The prevalence of subclinical endometritis and intrauterine infections in repeat breeder cows. *Theriogenology* 83:1249-1253.

Potter TJ, Guitian J, Fishwick J, Gordon PJ and Sheldon IM. 2010. Risk factors for clinical endometritis in postpartum dairy cattle. *Theriogenology* 74:127-134.

Prasad JK, Saxena MS, Prasad S and Singh GK. 2009. Comparative efficacy of *Escherichia coli* lipopolysaccharide, oyster glycogen and enrofloxacin on uterine defense mechanism and fertility in crossbred cows with endometritis. *Indian J Anim Sci* 79:1111-1115.

Prescott J F. 2000. Beta-lactam antibiotics: Penam penicillins. Pages 105–133 in *Antimicrobial Therapy in Veterinary Medicine*. 3rd ed. J. F. Prescott, J. D. Baggot, and R. D. Walker, ed. Iowa State University Press, Ames.

Priedkains J. 1987. Female reproductive system. In: Dellman, H.D., Brown, E.M. (Eds.), *Textbook of Veterinary Histology*, Lea & Febiger, Philadelphia, pp. 332-333.

Prunner I, Pothmann H, Wagener K, Giuliodori M, Huber J, Ehling-Schulz M and Drillich M. 2014a. Dynamics of bacteriologic and cytologic changes in the uterus of postpartum dairy cows. *Theriogenology* 82:1316-1322.

Prunner I, Wagener K, Pothmann H, Ehling-Schulz M and Drillich M. 2014b. Risk factors for uterine diseases on small-and medium-sized dairy farms determined by clinical, bacteriological, and cytological examinations. *Theriogenology* 82:857-865.

Purohit GN 2008. Recent developments in the diagnosis and therapy of repeat breeding cows and buffaloes. CAB Reviews: *Perspectives in Agriculture, Veterinary Science, Nutrition and Natural Resources*, **3**(062):1-34.

Purohit GN, Dholpuria S, Yadav S, Barolia Y and Kumar P. 2013. Diagnosis and treatment of clinical and subclinical endometritis-A clinical study in 30 cows. *Intas Polivet* 14:31-33.

Purohit GN, Ruhil S and Khichar V. 2015. Postpartum endometritis in dairy cows: current status of diagnosis, therapy and prevention. *Theriogenology Insight* 5:1-23

Pyorala S. 2000. Hirvonen's Thesis on Acute Phase Response in Dairy Cattle. Finnland: University of Helsinki.

Quereda JJ, Barba M, Mocé ML, Gomis J, Jiménez-Trigos E, García-Muñoz, Á, Gómez-Martín Á, González-Torres P, Carbonetto B and García-Roselló E. 2020. Vaginal microbiota changes during estrous cycle in dairy heifers. *Front Vet Sci* 7:p 371.

Quintal-Franco JA, Kojima FN, Melvin EJ, Lindsey BR, Zanella E, Fike KE, Wehrman ME, Clopton DT and Kinder JE. 1999. Corpus luteum development and function in cattle with episodic release of luteinizing hormone pulses inhibited in the follicular and early luteal phases of the estrous cycle. *Biol Reprod* 61:921-926.

Quintela LA, Becerra JJ and Herradón PG. 2016. Essential Guides on Cattle Farming: Uterine Diseases. *Servet: Zaragoza*; 69 p.

Radoslavov V and Savova-Burdarvo S. 1982. Occurrence of bacterial contamination, clinical symptoms and histopathlogical changes in the genital organs of cows after calving. *Vet Med Nauki* 19:76-85.

Raheel IAER, Hassan WH, Salem SSR and Salam HSH. 2020. Biofilm forming potentiality of Escherichia coli isolated from bovine endometritis and their antibiotic resistance profiles. *J Adv Vet Anim Res* 7: p 442.

Rahi S, Gupta HP, Prasad S and Baithalu RK. 2013. Phytotherapy for endometritis and subsequent conception rate in repeat breeding crossbreed cows. *Indian J Anim Reprod* 34:9-12.

Rahi S. 2011. Immunotherapeutic Effect of ashwagandha and garlic on endometritis in repeat bredding cows. M.V.Sc Thesis (Veterinary Gynaecology and Obstetrics), G.B. Pant University of Agricultural Sciences and Technology, Pantnagar.

Rahman MM and Lecchi C. 2010. Lipopolysaccharide binding protein: Local expression in bovine extrahepatic tissues. *Vet Immunol Immunopathol* 137:28-35.

Raja S, Devanathan TG, Kulasekar K, Pazhanivel N and Balachandran C. 2012. Whiteside test and endometrial biopsy for diagnosis of endometritis in repeat breeding cows. *Indian J Anim Reprod* 33:56-58.

Ramadin AA, Johnson GL and Lewis GS. 1997. Regulation of uterine immune function during the oestrous cycle and in response to infectious bacteria. *J Anim Sci* 75:1621-1632.

Ranasinghe RMSBK, Nakao T, Yamada K, Koike K, Hayashi A and Dematawewa CMB. 2011. Characteristics of prolonged luteal phase identified by milk progesterone concentrations and its effects on reproductive performance in Holstein cows. *J Dairy Sci* 94:116-127.

Rathaur P, Raja W, Ramteke PW and John S.A. 2012. Turmeric: The golden spice of life. *Int J Pharm Sci Res* 3:1987-1994.

Raval SR, Panchal MT, Dhami AJ, Hadiya KK and Bhanderi BB. 2018. Bacterial isolates and their antibiogram from genital Discharges of puerperal and repeat breeding crossbred Cows. *Indian J Anim Reprod* 39:58-60.

Razin B, Bauminge S and Globerson A. 1978. Effect of prostaglandins on phagocytosis of sheep erythrocytes by mouse peritoneal macrophages. *J Ret End Soc* 23:237-242.

Reames PS, Hatler TB, Hayes SH, Ray DL and Silvia WJ. 2011. Differential regulation of estrous behavior and luteinizing hormone secretion by estradiol-17β in ovariectomized dairy cows. *Theriogenology* 75:233-240.

Ricketts SW and Mackintosh M E. 1987. Role of anaerobic bacteria in eq uine endometritis. *J Reprod Fert* 35:(Supplement): 343-351.

Ridker PM. 2003. Clinical application of C-reactive protein for cardiovascular disease detection and prevention. *Circulation* 107:363-369.

Righter HF, Mercer HD, Kline DA and Carter GG. 1975. Absorption of antibacterial agents by the bovine involuting uterus. *Canadian Vet J* 16:10-15.

Risco C, Archbald L, Elliott J, Tran T and Chavatte P. 1994. Effect of hormonal treatment on fertility in dairy cows with dystocia or retained fetal membranes at parturition. *J Dairy Sci* 94:2562-2569.

Roberts SJ. 1986. Veterinary Obstetrics and genital diseases (Theriogenology). Published by the author, Ithaca New York, 1971, 381-359; 496-506.

Rocksin A, Gerlach GF and Schwarz S. 2007. The implementation of the broth microdilution method to determine bacterial susceptibility to antimicrobial agents. *Berl Munch Tierarztl Wochenschr* 120:42-49.

Rodrigues N, Kästle J, Coutinho T, Amorim A, Campos G, Santos V Marques LM, Timenetsky J and de Farias ST. 2015. Qualitative analysis of the vaginal microbiota of healthy cattle and cattle with genital-tract. *GMR* 14: 6518-6528.

Rodriguez I G, Gonzales HE and Gallego MIM. 1991. Correlation de la morfologia endometrial conaislamentos bacteriologicos em vacas infertiles de la sabana de Bogota. *Revista ICA: Publicacion Cientifica del Instituto Colombiano Agropecuari* 26:275-286.

Roelofs JB, Van Eerdenburg FJCM, Soede NM and Kemp B. 2005. Various behavioral signs of estrous and their relationship with time of ovulation in dairy cattle. *Theriogenology* 63:1366-1377.

Rogan D, Fumoso E, Rodryguez E, Wade J and Sanchez Bruni SF. 2010. Use of a mycobacterial cell wall extract (MCWE) in susceptible mares to clear experimentally induced endometritis with Strep. zooepidermicus. *J Equine Vet Sci* 27:112-117.

Roland L, Drillich M and Iwersen M. 2014. Hematology as a diagnostic tool in bovine medicine. *J Vet Diagn Invest* 26:592-598.

Romaniukowa K. 1984. Phagocytosis in the uterus. In: *Proceedings of the 10th International Congress on Animal Reproduction and AI.* Urbana, Champaign, IV: XI, pp. 30–33.

Roncada P, Ermini L, Schleuning A, Stracciari GL and Strocchia A. 2000. Pharmacokinetics and residual behaviour in milk of oxytetracycline in cows following administration of uterine pessaries. *J Vet Pharmacol Ther* 23:281-285.

Roszel J and Freeman K. 1988. Equine endometrial cytology. *Vet Clin North Am Equine Pract* 4:247-262.

Roth JA and Kaeberle ML. 1981. Effect of *in vivo* dexamethasone administration on in vitro bovine polymorphonuclear leukocyte function and lymphocyte blastogenesis. *Am J Vet Res* 43:412-416.

Roth JA and Kaeberle ML. 1985. *In vivo* effect of ascorbic acid on neutrophil function in healthy and dexamethasone-treated cattle. *Am J Vet Res* 46:2434-2436.

Roth JA, Kaeberle ML, Appell LH and Hachreiner RF. 1983. Association of increased estradiol and progesterone blood values with altered bovine polymorphonuclear leucocyte function. *Am J Vet Res* 44:247-253.

Rother R P, Bell L, Hillmen P and Gladwin MT. 2005. The clinical sequelae of intravascular hemolysis and extracellular plasma hemoglobin: a novel mechanism of human disease. *J Am Med Assoc* 293:1653-1662.

Runciman DJ, Anderson GA, Malmo J and Davis GM. 2008a. Use of postpartum vaginoscopic (visual vaginal) examination of dairy cows for the diagnosis of endometritis and the association of endometritis with reduced reproductive performance. *Aust Vet J* 86:205-213.

Runyoro DK, Matee MI, Ngassapa OD, Joseph CC and Mbwambo ZH. 2006. Screening of Tanzanian medicinal plants for anti-Candida activity. *BMC Complement Altern Med* 6, 11.

Sabedra DA. 2012. Thesis on Serum Haptoglobin as an Indicator for Calving Difficulties and Postpartal Diseases in Transition Dairy Cows. An Undergraduate Thesis Submitted to Oregon State University.

Sachi S, Ferdous J, Sikder MH, Azizul Karim Hussani SM. 2019. Antibiotic residues in milk: Past, present, and future. *J Adv Vet Anim Res*. 6:315-332.

Sadanandan B, Prerna L, Humtsoe H and Mishra A. 2014. Antibacterial activity of garlic against *Bacillus subti-lis*. *Int Rev Appl Biotechnol Biochem* 2:107-119.

Sagartz JW and Hardenbrook HJ. 1971. A clinical, bacteriologic, and histologic survey of infertile cows. *J Am Vet Med Assoc* 158:619-622.

Salasel B, Mokhtari A and Taktaz T. 2010. Prevalence, risk factors for and impact of subclinical endometritis in repeat breeder dairy cows. *Theriogenology* 74:1271-1278.

Salphale GB, Kadu MM, Fasihddin M and Kadu MS. 1993. Study of some physical properties of estrous cervical mucus in synchronized animals and Repeat breeder cross bred cows with reference to fertility. *Indian J Anim Reprod* 14:77-78.

Sannmann I and Heuwieser W. 2015. Technical note: Intraobserver, interobserver, and test-retest reliabilities of an assessment of vaginal discharge from cows with and without acute puerperal metritis. *J Dairy Sci* 98:5460-5466.

Sannmann I, Arlt S and Heuwieser W. 2012. A critical evaluation of diagnostic methods used to identify dairy cows with acute postpartum metritis in the current literature. *J Dairy Res* 79:436-444.

Sannmann I, Burfeind O, Suthar V, Bos A, Bruins M and Heuwieser W. 2013a. Technical note: Evaluation of odor from vaginal discharge of cows in the first 10 days after calving by olfactory cognition and an electronic device. *J Dairy Sci* 96:5773-5779.

Sannmann I, Burfeind O, Voigtsberger R and Heuwieser W. 2013b. Comparison of two monitoring and treatment strategies for cows with acute puerperal metritis. *Theriogenology* 79:961-969.

Santos NR, Roman HB and Gilbert RO. 2006. The use of leukocyte esterase reagent strips for diagnosis of subclinical endometritis in dairy cows. *Theriogenology* 66:666-667.

Santos TM and Bicalho RC. 2012. Diversity and succession of bacterial communities in the uterine fluid of postpartum metritic, endometritic and healthy dairy cows. *PloS One* 7: p 53048.

Santos TM, Caixeta LS, Machado VS, Rauf AK, Gilbert RO and Bicalho RC. 2010a. Antimicrobial resistance and presence of virulence factor genes in Arcanobacterium pyogenes isolated from the uterus of postpartum dairy cows. *Vet Microbiol* 145:84-89.

Santos TM, Gilbert RO, Bicalho RC. 2011. Metagenomic analysis of the uterine bacterial microbiota in healthy and metritic postpartum dairy cows. *J Dairy Sci* 94:291-302.

Santos TM, Gilbert RO, Caixeta LS, Machado VS, Teixeira LM and Bicalho RC. 2010b. Susceptibility of *Escherichia coli* isolated from uteri of postpartum dairy cows to antibiotic and environmental bacteriophages. Part II: in vitro antimicrobial activity evaluation of a bacteriophage cocktail and several antibiotics. *J Dairy Sci* 93:105-114.

Saremi B, Al-Dawood A, Winand S, Müller U, Pappritz J, von Soosten D, Rehage J, Dänicke S, Häussler S, Mielenz M. and Sauerwein H. 2012. Bovine haptoglobin as an adipokine: Serum concentrations and tissue expression in dairy cows receiving a conjugated linoleic acids supplement throughout lactation. *Vet Immunol Immunopath* 146:201-211.

Sarkar P, Kumar H, Rawat M, Varshney VP, Goswami TK, Yadav MC and Srivastava SK. 2006. Effect of administration of garlic extract and PGF2α on hor-monal changes and recovery in endometritis cows. *Asian Aus J Anim Sci* 19:964-969.

Sarmah AK, Meyer MT, Boxall AB. 2006. A global perspective on the use, sales, exposure pathways, occurrence, fate and effects of veteri-nary antibiotics (VAs) in the environment. *Chemosphere* 65:725-759.

Sartori R and Barros CM. 2011. Reproductive cycles in *Bos indicus* cattle. *Anim Reprod Sci* 124:244-250.

Sartori R, Haughian JM, Shaver RD, Rosa GJM and Wiltbank MC. 2004. Comparison of ovarian function and circulating steroids in estrous cycles of Holstein heifers and lactating cows. *J Dairy Sci* 87:905-920.

Satheshkumar S and Punniamurthy N. 2007. Sub-clinical uterine, infections in repeat breeder cows. *Indian Vet J* 84:654-655.

Sattler T and Furll M. 2004. Creatine kinase and aspartate aminotransferase in cows as indicators for endometritis. *J Vet Med A* 51:132-137.

Saumande J and Humblot P. 2005. The variability in the interval between estrus and ovulation in cattle and its determinants. *Anim Reprod Sci* 85:171-182.

Savitha BH, Praseeda R and Ajitkumar G. 2005. Whiteside test for subclinial endometritis in cattle. *J Vet Anim Sci* 38:153-154.

Sawant AA, Sordillo LM and Jayarao BM. 2005. A survey on antibiotic usage in dairy herds in Pennsylvania. *J Dairy Sci* 88:2991-2999.

Schaer DJ, Schaer CA, Buehler PW, Boykins RA, Schoedon G, Alayash AI and Schaffner A. 2006. CD163 is the macrophage scavenger receptor for native and chemically modified hemoglobins in the absence of haptoglobin. *Blood* 107:373-380.

Schalm OW, Carroll EJ and Jain NC. 1971. *Bovine Mastitis*, Lea and Febiger, Philadelphia.

Schuetz P, Albrich W and Mueller B. 2011. Procalcitonin for diagnosis of infection and guide to antibiotic decisions: past, present and future. *BMC Med* 9:107.

Scrollavezza P, Ablondi M, Pogliacomi B, Guareschi D, Dall'Aglio R, Poldi R and Pezzoli G., 1997. Ozone treatment in mastitis, metritis and retention of fetal membranes in the cows. *Atti 2 Intern Symp Ozone Application*, Havana, Cuba.

Segerson EC, Godfrey RW, Randel RD and Gunsett FC, 1986. Immunosuppressive activity of uterine luminal protein from steroid-treated ovariectomised heifers. *Theriogenology* 25:639-651.

Segerson EC, Libby DW, Getz WR and Randel RD. 1984. Immunosuppressive effect of uterine secretory protein from Angus and Brahman cows upon lymphocytes in *vitro*. *J Anim Sci* 59:1047-1059.

Seguin BE, Morrow DA and Louis TM. 1974. Luteolysis, luteostasis and the effect of prostaglandin F2 alpha in cows with endometrial irritation with Lugol's. *Am J Vet Res* 35:57-59.

Senior J, Sangha R, Baxter GS, Marshall K and Clayton JK. 1992. *In vitro* characterization of prostanoid FP- DP-, IP- and TP-receptors on the non-pregnant human myometrium. *Br J Pharmacol* 107:215-221.

Setkus A, Galdikas A-J, Kancleris Z-A, Olekas A, Senuliene D, Strazdiene V, Rimdelka R and Bagdonas R. 2006. Featuring of bacterial contamination of wounds by dynamic response of SnO2 gas sensor array. *Sensor Actuat BChem* 115:412-420.

Shacter E, Arzadon G and Williams J. 1992. Elevation of interleukin 6 in response to a chronic inflammatory stimulus in mice: Inhibition by indomethacin. *Blood* 80:194-202.

Shameli K, Ahmad M, Bin Zamanian A, Sangpour P, Shabanzadeh P and Abdollahi Y. 2012. Green biosynthesis of silver nanoparticles using *Curcuma longa* tuber powder. *Int J Nanomedicin* 7:5603-5610.

Sharma P, Srivastava S, Kumar R and Singh VB. 2018. Phytotherpary: An alternative low cost therapeutic management of endometritis in dairy animals. *Int J Curr Microbiol App Sci* Special Issue-7: 4581-4591.

Sharma SK, Sharma H, Dhami AJ and Bhong CD. 2008. Physico-microbial properties of cervico-vaginal mucus and its antibiotic sensitivity pattern in repeat breeding buffaloes. *Indian J Anim Reprod* 29:19-26.

Sharon N. 2006. Carbohydrates as future anti adhesion drugs for infectious diseases. *Biochimica et Biophysica Acta* (BBA) General Subjects, 1760, 527-537.

Sheldon IM and Dobson H. 2004. Postpartum uterine health in cattle. *Anim Reprod Sci* 82-83:295-306.

Sheldon IM and Noakes DE. 1998. Comparison of three treatments for bovine endometritis. *Vet Rec* 142:575-579.

Sheldon IM, Bushnell M, Montgomery , Rycroft AN. 2004a. Minimum inhibitory concentration of some antimicrobial drugs against bacteria causing uterine infections in cattle. *Vet Rec* 155:383-387.

Sheldon IM, Cronin J, Goetze L, Donofrio G and Schuberth HJ. 2009a. Defining postpartum uterine disease and the mechanisms of infection and immunity in the female reproductive tract in cattle. *Biol Reprod* 81:1025-1032.

Sheldon IM, Cronin JG and Bromfield JJ. 2019. Tolerance and innate immunity shape the development of postpartum uterine disease and the impact of endometritis in dairy cattle. *Annu Rev Anim Biosci* 7: p 361.

Sheldon IM, Lewis GS, LeBlanc S and Gilbert RO. 2006. Defining postpartum uterine disease in cattle. *Theriogenology* 65:1516-1530.

Sheldon IM, Molinari PC, Ormsby TJ and Bromfield JJ. 2020. Preventing postpartum uterine disease in dairy cattle depends on avoiding, tolerating and resisting pathogenic bacteria. *Theriogenology* 150:158-165.

Sheldon IM, Noakes DE, Rycroft A and Dobson H. 2001. Acute phase protein responses to uterine bacterial contamination in cattle after calving. *Vet Rec* 148:172-175.

Sheldon IM, Noakes DE, Rycroft AN and Dobson H. 2002a. Effect of postpartum manual examination of the vagina on uterine bacterial contamination in cows. *Vet Rec* 151:531-534.

Sheldon IM, Noakes DE, Rycroft AN and Dobson H. 2004c. Effect of intrauterine administration of estradiol on postpartum uterine bacterial infection in cattle. *Anim Reprod Sci* 8:13-23.

Sheldon IM, Noakes DE, Rycroft AN, Pfeiffer DU and Dobson H. 2002b. Influence of uterine bacterial contamination after parturition on ovarian dominant follicle selection and follicle growth and function in cattle. *Reproduction-Cambridge* 123:837-845.

Sheldon IM, Price SB, Cronin J, Gilbert RO and Gadsby JE. 2009b. Mechanisms of infertility associated with clinical and subclinical endometritis in high producing dairy cattle. *Reprod Dom Anim.* 44(Suppl. 3):1-9.

Sheldon IM, Rycroft AN and Zhou C. 2004b. Association between postpartum pyrexia and uterine bacterial infection in dairy cattle. *Vet Rec* 154:289-293.

Sheldon IM, Rycroft AN, Dogan B, Craven M, Bromfield JJ, Chandler A, Roberts MH, Price SB, Gilbert RO and Simpson KW. 2010. Specific strains of *Escherichia coli* are pathogenic for the endometrium of cattle and cause pelvic inflammatory disease in cattle and mice. *PloS One* 5(2): p 9192.

Sheldon IM, Williams EJ, Miller ANA, Nash DM and Herath S. 2008. Uterine diseases in cattle after parturition. *Vet J* 176:115-121.

Sheldon IM. 2004. The postpartum uterus. *Vet Clin N Am Food Anim Prac* 20:569-591.

Sheldon IM. 2015. Physiology and endocrinology symposium: Uterine infection: Linking infection and innate immunity with infertility in the high-producing dairy cow. *J Anim Sci* 93:2021-2033.

Shimizu T, Miyauchi K, Shirasuna K, Bollwein H, Magata F, Murayama C and Miyamoto A. 2012. Effects of lipopolysaccharide (LPS) and peptidoglycan (PGN) on estradiol production in bovine granulosa cells from small and large follicles. *In Vitro Toxicol* 26:1134-1142.

Shukla SP and Pandit RK. 1989. Incidence of repeat breeding and its remedial measures in Gir cows and their crosses. *Indian Vet J* 66:626-630.

Shukla SP and Sharma R D. 2005. Bacteriological studies on the uterine biopsy and conception rate foil owing treatment in repeat breeding crossbred cows. *Indian J Anim Reprod* 26:17-19.

Sicsic R, Goshen T, Dutta R, Kedem-Vaanunu N, Kaplan-Shabtai V, Pasternak Z, Gottlieb Y, Shpigel NY and Raz T. 2018. Microbial communities and inflammatory response in the endometrium differ between normal and metritis dairy cows at 5-10 days post-partum. *Vet Res* 49:77.

Sikrodia R, Chhabra D, Mahor SS, Audarya SD, Gangil R and Sharda R. 2021. Isolation of Staphylococcus from repeat breeder crossbred cattle and their antibiotic sensitivity pattern. *J Entomol Zool Stud* 9:2158-2160.

Silvestre FT, Carvalho TS, Francisco N, Santos JE, Staples CR, Jenkins TC and Thatcher WW. 2011. Effects of differential supplementation of fatty acids during the peripartum and breeding periods of Holstein cows: I. Uterine and metabolic responses, reproduction, and lactation. *J Dairy Sci* 94:189-204.

Simon L, Gauvin F, Amre DK, Saint-Louis P and Jacques Lacroix J. 2004. Serum procalcitonin and C-reactive protein levels as markers of bacterial infection: a systematic review and meta-analysis. *Clin Infect Dis* 39:206-217.

Singh AK, Nigam JM and Sharma DN. 1983. Histological observation on endometrial gland of cows and buffaloes in relation to fertility. *Trop Vet Anim Sci Res* 1:37-41.

Singh J, Nanda AS, Dhaliwal GS and Pangaonkar GR. 2003. Treatment of bacterial endometritis in crossbred cows using intrauterine oy ster glycogen, a non specific immunomodulators. *Ind J Anim Sci* 73:844-847.

Singh J, Sidhu SS, Dhaliwal GS, Pangaonkar GR, Nanda AS and Grewal AS. 2000. Effectiveness of lipopolysaccharide as an intrauterine immunomodulator in curing bacterial endometritis in repeat breeding cross-bred cows. *Anim Reprod Sci* 59:159-166.

Singh MK, Jayachandran C, Roy GP and Banerjee NC. 1988. Pharmacokinetics and distribution of sulfadimidine in plasma, milk and uterine fluids of female buffaloes. *Vet Res Commun* 12:41-46.

Singh S. 2016. Phytotherapeutic Measures for Endometritis in Crossbred Cows. M.V.Sc. Thesis (Veterinary Gynaecology and Obstetrics). Nanaji Deshmukh Veterinary Science University, Jabalpur.

Singla P, Singh J, Dhaliwal GS and Kumar A. 2004. Effect of post AI intrauterine treatment with E. coli lipopolysaccharide and autologous serum on characteristics of cervico-vaginal mucous in subclinical endometritic cows. *Indian J Anim Reprod* 25:55-57.

Sinha RN, Sinha AK, Alam M and Singh B. 1994. Concentration of gentamicin in endometrial tissues subsequent to muscular and uterine routes of administration in cows. *Indian J Anim Sci* 15:29-31.

Siregar TN, Armansyah T, Panjaitan B, Gholib G, Herrialfian H, Sutriana A, Abidin Z, Reynaldi MA, Razak F, Artaliani Y and Yuswar Y. 2019. Changes in cervical mucus as an indicator of fertility in Aceh Cattle. *Adv Anim Vet Sci* 7:306-314.

Skhervan O. 1954. The intrauterine distribution and absorption of medicaments. *Proceedings VII Nord Vet Cong* pp. 236-245. (Cited by Masera *et al.* 1980a).

Skidmore AL. 2015. Compendium of Animal Reproduction. *Intervet Internantional.* 11th edition.

Skinner JG, Brown RA and Roberts, L. 1991. Bovine haptoglobin response in clinically defined field conditions. *Vet Rec* 128:147-149.

Skorikov VN, Kuchmenko TA, Mikhalev1 V I and Umarkhanov R U. 2021.The use of the electronic nose device for the diagnosis of postpartum metritis in cows *IOP Conf. Ser. Earth Environ Sci* 640 072032

Slama H, Vaillancourt D and Goff AK. 1991. Pathophysiology of the puerperal period: Relationship between prostaglandin E2 (PGE2) and uterine involution in the cow. *Theriogenology* 36:1071-1090.

Slocombe LL and Colditz IG. 2005. Evaluating The Stress of Production in Cattle Using Haptoglobin. *5th International Colloquium on Animal Acute Phase Proteins*. Dublin.

Smith BI and Risco CA. 2005. Management of periparturient disorders in dairy cattle. *Vet Clin North Am Food Anim Pract* 21:503-521.

Smith BI, Donovan GA, Risco C, Littell R, Young C, Stanker LH and Elliott J. 1998b. Comparison of various antibiotic treatments for cows diagnosed with toxic puerperal metritis. *J Dairy Sci* 81:1555-1562.

Smith BI, Donovan GA, Risco CA, Young CR and Stanker LH. 1998a. Serum haptoglobin concentrations in Holstein dairy cattle with toxic puerperal metritis. *Vet Res* 142:83-85.

Sood P and Ahuja AK. 2023. Prevalence of single and multiple reproductive disorders in Jersey crossbred cows of sub-tropical North-Western Himalayas. *Indian J Anim Sci* 93:697-700.

Sood P, Ahuja A, Mankotia R, Nevi V, Pandey AK and Kumar S. 2024. Bovine Endometrium Cytotaping Catheter. BIS:2024, Manak Bhawan, New Delhi (under review).

Sood P, Kumar NV, Singh M and Pathania N. 2009. Prevalence and certain characteristics of mid-cyclic estrus in crossbred cows. *Vet Arh* 79:143-149.

Sood P, Nanda AS and Srivastava A K. 1999a. Disposition kinetics and uterine tissue level of neomycin in buffaloes. *Indian Vet J* 76:1071-1073.

Sood P, Nanda AS and Srivastava A K. 1999b. Disposition kinetics, uterine tissue concentration and dosage regimen of neomycin in dystocia-affected buffaloes (*Bubalus bubalis*) following its single intravenous administration. *Buffalo Bull* 18:32-36.

Sood P, Nanda AS and Srivastava A K. 2004. Disposition kinetics, dosage regimen and nephrotoxic evaluation of neomycin following intrauterine infusion in endometritis buffaloes. *Indian Vet J* 81:144-147.

Sood P, Nanda AS and Srivastava A K. 1998. Plasma levels and pharmacokinetics of neomycin in buffaloes (*Bubalus bubalis*) following its single intrauterine infusion. *Indian J Anim Sci* 68:317-319.

Sood P, Nanda AS and Srivastava AK. 2002b. *In vitro* erythrocytic penetration of neomycin in reproductively healthy and endometritis buffaloes. *J Vet Pharmacol and Toxicol* 2:104-106.

Sood P, Sarma HD, Kadwad VB and Kumar S. 2022. Effect of ovarian follicular wave pattern and endocrine characteristics on pregnancy outcome in cows. *Reprod Dom Anim* 57:321-332.

Sood P, Thakur SD, Sarvesh, Dogra V, Vasishta NK, Singh M and Gandhotra VK. 2012. Evaluation of efficacy, intrauterine safety and storage conditions of Lugol's iodine. *Indian J Anim Reprod* 33:71-73.

Sood P, Verma S, Katoch RC, Singh M and Vasishta NK. 2003. Impact of microbial panorama on the therapeutic efficacy of single injection of PGF_2 alpha in cows with clinical endometritis. *Indian J Anim Sci* 73:68-69.

Sood P, Verma S, Mahajan A, Singh M, Vashishta NK and Katoch RC. 2002c. Microbiological profile of genital discharge in cycling endometritic cows of Palam valley of Himachal Pradesh. *Indian J Anim Sci* 72:251-252.

Sood P, Verma S, Vasishta NK, Singh M and Katoch RC. 2000. Factors affecting therapeutic efficacy of PGF_2 Alpha, uterine infusion of Lugol's iodine during or one day after estrus in management of clinical endometritis in cycling cows. *Himachal Vet J* 4:1-6.

Sood P, Zachut M, Dube H and Moallem U. 2015. Behavioral and hormonal pattern of repeat breeder cows around estrus. *Reproduction* 149:545-554.

Sood P. 1995. Disposition of neomycin in dystocia and endometritis affected buffaloes. M.V.Sc Thesis. Punjab Agricultural University, Ludhiana.

Sood P. 2002a. Therapeutic strategies involving antimicrobial treatment of the uterus in large animals. Proceedings Orientation Course on "Industrial Toxicology" from 9th to 30th September, 2002. P.A.U., Ludhiana.

Srivastava AK and Malik JK. 1987. Studies on distribution of deacetylmonoxime in body fluids of buffalo calves following its single intravenous administration. *J Remount Vet Corps* 26:137-142.

Starbuck G. 1999. The importance of progesterone during early pregnancy in the dairy cow. *Cattle Pract* 7:397-399.

Steffan J, Agrie M, Adriamanga S and Thibier M. 1984. Treatment of metritis with antibiotics or prostaglandin F2α and influence of ovarian cyclicity in dairy cows. *Am J Vet Res* 45:1090-1094.

Stevenson JS, Lamb GC, Kobayashi Y and Hoffman DP. 1998. Luteolysis during two stages of the estrous cycle: subsequent endocrine profiles associated with radiotelemetrically detected estrus in heifers. *J Dairy Sci* 81:2897-2903.

Stevenson JS. 1997. Clinical reproductive physiology of the cow. Pages 257–267 in *Current Therapy in Large Animal Theriogenology*. R. S. Youngquist, ed. W.B. Saunders Co., Philadelphia, PA.

Stojkov J, von Keyserlingk MA, Marchant-Forde JN and Weary DM. 2015. Assessment of visceral pain associated with metritis in dairy cows. *J Dairy Sci* 98:5352-5361.

Stolla R and Himmer B. 1980. Probleme bei rektaler Untersuhung der Ovarien. *Prak Tierarzt* 61:9-12.

Strowitzki T, Germeyer A, Popovici R and von Wolff M. 2006. The human endometrium as a fertility-determining factor. *Hum Reprod Update* 12:617-630.

Strube K, Hühn R, Busch W and Werner E. 1991. Ein Phagozytosetest zur Einschätzung der lokalen Abwehrsituation bei Endometritistherapie unter besonderer Berücksichtigung des Uterofertileinsatzes beim Rind [A phagocytosis test for the evaluation of the local immunity status during endometritis therapy with special regard to the use of uterofertil in cattle]. *Dtsch Tierarztl Wochenschr* 98:230-234.

Struve K, Herzog K, Magata F, Piechotta M, Shirasuna K, Miyamoto A and Bollwein H. 2013. The effect of metritis on luteal function in dairy cows. *BMC Vet Res* 9:244.

Studer E and Morrow DA. 1978. Postpartum evaluation of bovine reproductive potential: Comparison of findings from genital tract examination per rectum, uterine culture and endometrial biopsy. *J Am Vet Med Assoc* 172:489-494.

Studer E. 1981. Evaluation of the postpartum reproductive tract in the bovine. In: *Proc. Ann. Mtg. Soc. For Theriogenology*, Spokane, Wa., 113-125.

Subandrio AL and Noakes DE. 1992. The influence of the stage of the bovine oestrous cycle on the chemotactic stimulus of oyster glycogen to intrauterine neutrophils. *Br Vet J* 148:163-165.

Subandrio AL and Noakes DE. 1997. Neutrophil migration into the uterine lumen of the cow: The influence of endogenous sex steroid hormones using two intrauterine chemoattractants. *Theriogenology* 47:825-835.

Subandrio AL, Sheldon IM and Noakes DE. 2000. Peripheral and intrauterine neutrophil function in the cow: the influence of endogenous and exogenous sex steroid hormones. *Theriogenology* 53:1591-1608.

Sumiyoshi T, Tanaka T and Kamomae H. 2014. Relationships between the appearances and changes of estrous signs and the estradiol-17β peak, luteinizing hormone surge and ovulation during the periovulatory period in lactating dairy cows kept in tie-stalls. *J Reprod Dev* 60:106-114.

Supre K, Lommelen K and Meulemeester LD. 2014. Antimicrobial susceptibility and distribution of inhibition zone diameters of bovine mastitis pathogens in Flanders, Belgium. *Vet Microbiol* 171:374-381.

Suriyasathaporn W, Heuer C, Noordhuizen-Stassen EN and Schukken YH. 2000. Hyperketonemia and udder defense: A review. *Vet Res* 31:397-412.

Swidsinski A, Verstraelen H, Loening-Baucke V, Swidsinski S, Mendling W and Halwani Z. 2013 Presence of a polymicrobial endometrial biofilm in patients with bacterial vaginosis. *PLoS One* 8 e53997.

Szenci O. 2016. Recent possibilities for diagnosis and treatment of postparturient uterine diseases in dairy cow. *JFIV* 4.1:1-7.

Takahashi E, Uzuka Y, Tanabe S, Satoh M and Furuoka H. 2006. Serum amyloid A and haptoglobin levels in bovine. *J Vet Med* 69:321-323.

Talukder J, Srivastava A, Ray A and Lall R. 2018. Treatment of infectious endometritis with a novel protein, VPI-O22, in Cows. *FASEB Journal* 32:882.12-882.12.

Targowski SP. 1984. Immunologically induced uterine inflammation. In: *Proceedings of the 10th International Congress on Animal Reproduction and AI. Urbana, Champaign,* VI, pp. 473-475.

Taylor VJ, Beever DE, Bryant MJ and Wathes DC. 2003. Metabolic profiles and progesterone cycles in first lactation dairy cows. *Theriogenology* 59:1661-1677.

Tenhagen BA, Heuwieser W. 1999. Comparison of a conventional reproductive management program based on rectal palpation and uterine treatment of endometritis with a strategic prostaglandin F2a-program. *J Vet Med Assoc* 46:167-176.

Theilgaard-Monch K, Jacobsen LC, Nielsen MJ, Rasmussen T, Udby, Gharib M, Arkwright PD, Gombart A. F, Calafat J, Moestrup SK, Porse BT and Borregaard N. 2006. Haptoglobin is synthesized during granulocyte differentiation, stored in specific granules, and released by neutrophils in response to activation. *Blood* 108:353-61.

Tizard IR. 1996. *Veterinary Immunology*: An Introduction, fifth ed. W.B. Saunders Co., Philadelphia, PA, pp. 32-87.

Tombácz K, Peters LM, Chang YM, Steele M, Werling D and Gibson AJ. 2019. Short communication: Pegbovigrastim treatment *in vivo* does not affect granulocyte ability to migrate to endometrial cells and kill bacteria in vitro in healthy cows. *J Dairy Sci* 102:9389-9395.

Toutain PL Bousquet-Mélou A Damborg P Ferran AA Mevius D Pelligand L Veldman KT Lees P. 2017. En route towards European clinical breakpoints for veterinary antimicrobial susceptibility testing: A position paper explaining the VetCAST approach. *Front Microbiol* 8:2344.

Tragesser LA, Wittum TE, Funk JA, Winokur PL and Rajala-Schultz PJ. 2006. Association between ceftiofur use and isolation of Escherichia coli with reduced susceptibility to ceftriaxone from fecal samples of dairy cows. *Am J Vet Res* 67:1696-1700.

Tras B and Uney K, 2011. The rational use of antibacterial drugs in cows and failure causes in the treatment with antibacterial drugs. *Turk Vet Hek Bir Derg* 34:81-100.

Tras B, Dinc DA and Uney K. 2014. The effect of N-acetylcysteine on the treatment of clinical endometritis and pregnancy rate in dairy cows. *Eurasian J Vet Sci* 30:133-137.

Trimbos J and Arentz N. 1985. The efficiency of the cytobrush versus the cotton swab in the collection of endocervical cells in cervical smears. *Acta Cytol* 30:261-263.

Troedsson MHT, Liu IKM, Ing M and Pascoe J. 1995. Smooth muscle electrical activity in the oviduct and the effect of oxytocin, prostaglandin F2α and prostaglandin E2 on the myometrium and the oviduct of the cycling mare. *Biol Reprod Mono 1: Equine Reprod.* VI: 475-488.

Troedsson MHT. 1999. Uterine clearance and resistance to persistent endometritis in the mare. *Theriogenology* 52:461-471.

Tsai J, Yang F, Wang C, Fang T, Lee R and Hsu B. 2010. Effect of intravenous N-acetylcysteine on plasma total homocysteine and inflammatory cytokines during high flux heomodialysis. *Tzu-Chi Medical J* 22:91-95.

Tseng CF, Lin CC, Huang HY, Liu HC and Mao SJ. 2004. Antioxidant role of human haptoglobin. *Proteomics* 4: 2221-2228.

Tsiligianni T, Amiridis GS, Dovolou E, Menegatos I, Chadio S, Rizos D and Gutierrez Adan A. 2011. Association between physical properties of cervical mucus and ovulation rate in superovulated cows. *Canadian J Vet Res* 75:248-253.

Tsuru S, Oguchi M, Kitani H, Ikeda M, Seno M, Miake S, Zinnaka Y and Nomoto K. 1984. Relationship between bactericidal and phagocytic activities of peritoneal macrophages induced by irritants. *J Clin Lab Immunol* 13:81-84.

Uchida E, Katoh N, Bell AW and Overton TR. 2006. Appearance of haptoglobin in serum from cows at parturition. *J Vet Med Sci* 55:893-894.

Ulug U, Goldman S, Ben-Shlomo I and Shalev E. 2001. Matrix metalloproteinase (MMP)-2 and MMP-9 and their inhibitor, TIMP-1, in human term decidua and fetal membranes: The effect of prostaglandin F(2alpha) and indomethacin. *Mol Hum Reprod* 7:1187-1193.

Umadevi M, Rajeswari R, Sharmila RC, Selvavenkadesh S, Pushpa R, Sampath-Kumar KP and Bhowmik D. 2012. Traditional and medicinal use of Withania somnifera. *Pharm Innov* 1:102-110.

Uppal RP, Gupta RC and Chaudhary SK. 1991. Pharmacokinetic profile of trimethoprim plus sulfamethoxazole following intrauterine administration in normal and metritis condition in buffaloes. *Indian Vet Med J* 15:78-82.

Valergakis GE, Oikonomou G, Arsenos G and Banos G. 2011. Phenotypic association between energy balance indicators and reproductive performance in primiparous Holstein cows. *Vet Rec* 168:189.

van der Burgt G, Clark W and Knight R. 2007. Cattle fertility problems and Histophilussomni. *Vet Rec* 160: p 600.

Van Gorp H, Delputte PL. and Nauwynck HJ. 2010. Scavenger receptor CD163, a Jack-ofall- trades and potential target for cell-directed therapy. *Mol Immunol* 47:1650-1660.

Van Sun RJ. 2004. *Metabolic profiling and health risk in transition cows.* In: *Proc. Am. Assoc. Bov. Pract.* 37:212-213.

Vandeplassche M and Bouters R. 1976. Puerperal metritis in the bovine. In: *Proceedings of the 4th International Congress on Animal Reproduction and AI.* Krakow, Poland, pp. 660-666.

Vandeplassche M and Bouters R. 1983. Phagocytosis in the blood and uterine exudate of mares, cows and sows. In: *Proceedings of the 3rd International Symposium on World Association for Veterinary Laboratory Diagnosis.* Ames, I, pp. 83-89.

Vandeplassche M. 1981. New comparative aspects of involution and puerperal metritis in mare, cow and sow. *Vet Med* 36:804-807.

Vandeplassche M. 1984. Stimulation and inhibition of phagocytosis in domestic animals. In: *Proceedings of the 10th International Congress on Animal Reproduction and AI. Urbana*, Champaign, III, pp. 475-477.

Vander Weilen AL and King GJ. 1984. Intraepithelial lymphocytes in the bovine uterus during the oestrous cycle and early gestation. *J Reprod Fertil* 70:457-462.

Vasconcelos JLM, Sartori R, Oliveira HN, Guenther JG and Wiltbank MC. 2001. Reduction in size of the ovulatory follicle reduces subsequent luteal size and pregnancy rate. *Theriogenology* 56:307-314.

Ventola CL. 2015. The antibiotic resistance crisis: part 1: causes and threats. *P T* 40:277-283.

Venugopal R. 1995. Intra uterine infusion of homologous plasma in the treatment of endometritis in cows. M.V.Sc. Thesis, College of Veterinary and Animal Sciences, Mannuthy, Thrissur.

Wade DE and Lewis GS. 1996. Exogenus prostaglandins F2α stimulate utero-ovarian release of PGF2α in sheep: A possible component of the leukocyte mechanism of action of PGF2α. *Dom Anim Endocrinol* 13:383-395.

Wagener K, Prunner I, Pothmann H, Drillich M and Ehling-Schulz M. 2015. Diversity and health status specific fluctuations of intrauterine microbial communities in postpartum dairy cows. *Vet Microbiol* 175:286-293.

Wagner C, Sauermann R and Joukhadar C. 2006. Principles of antibiotic penetration into abscess fluid. *Pharmacology* 78:1-10.

Wakayo BU, Brar PS, Prabhakar S and Arora AK. 2015. Evaluation of Prophylactic Cloprostenol and E. coli LPS Treatment Against Endometritis in Dairy Cows and Buffaloes. *J Reprod Infert* 6:17-21.

Waldron MR, Nonnecke BJ, Nishida T, Horst RL and Overton TR. 2003. Effect of lipopolysaccharide infusion on serum macromineral and vitamin D concentration in dairy cows. *J Dairy Sci* 86:3440-3446.

Walsh RB, Walton JS, Kelton DF, LeBlanc SJ, Leslie KE and Duffield TF. 2007. The effect of subclinical ketosis in early lactation on reproductive performance of postpartum dairy cows. *J Dairy Sci* 90:2788-2796.

Wang L, Bassiri M, Najafi R, Najafi K, Yang J, Khosrovi B, Hwong W, Barati E, Belisle B, Celeri C and Robson MC. 2007. Hypochlorous acid as a potential wound care agent: part I. Stabilized hypochlorous acid: a component of the inorganic armamentarium of innate immunity. *J Burns Wounds*. Apr 11;6:e5..

Ward A. 1985. A new treatment for infertility in mares. *Vet Med* 80:65-69.

Wathes DC, Cheng Z, Chowdhury W, Fenwick MA, Fitzpatrick R, Morris DG, Patton J and Murphy JJ. 2009. Negative energy balance alters global gene expression and immune responses in the uterus of postpartum dairy cows. *Physiol Genomics* 39:1-13.

Watson ED, Diehl NK and Evans JF. 1990. Antibody response in the bovine genital tract to intrauterine infusion of Actinomyces pyogenes. *Res Vet Sci* 48:70-75.

Watson ED. 1984. Plasma concentrations of PGFM in two cows with and in two cows without postpartum endometritis. *Vet Rec* 114:479-480.

Watson ED. 1985. Opsonising ability of bovine uterine secretions during the oestrous cycle. *Vet Rec* 117:274-275.

Weems CW, Weems YS and Randel RD. 2006. Prostaglandins and reproduction in female farm animals. *Vet J* 171:206-228.

Weinkauf C, Hachenberg S, Hiss S, Muller U and Sauerwein H. 2005. Haptoglobin in serum and milk of cows and its relation to udder health and metabolic disorders. *5th International Colloquium on Animal Acute Phase Proteins*, Dublin. 32.

Westermann S, Drillich M, Kaufmann TB, Madoz LV and Heuweiser W. 2010. A clinical approach to determine false positive findings of clinical endometritis by vaginoscopy by the use of uterine bacteriology and cytology in dairy cows. *Theriogenology* 72:1248-1255.

Whicher J, Bienvenu J and Monneret G. 2001. Procalcitonin as an acute phase marker. *Ann Clin Biochem* 38:483-93.

Whitacre M D. 1992. Intrauterine infusion in the postpartum dairy cow. *Vet Med (USA)* 87:376-381.

White AJ and Dobson H. 1990. Effect of prostaglandin F2α on the fertility of dairy cows after calving. *Vet Rec* 127:588-592.

Whiteside WH. 1939. Observations on a new test for the presence of mastitis in milk, *Can Pub Health J* 30:44.

Whitman WB, Coleman DC and Wiebe WJ. 1998. Prokaryotes: the unseen majority. *Proc Natl Acad Sci U S A* 95:6578-6583.

Whitmore HL and Archbald LF. 1977. Demonstration and quantitation of immunoglobulins in bovine serum, follicular fluid, and uterine and vaginal secretions with reference to bovine viral diarrhoea. *Am J Vet Res* 38:455-457.

Whitmore HL, Bernoco M, Liu IKM and Gustafsson BK. 1988. Effect of parturition on phagocytosis and chemotactic function in peripheral blood polymorphonuclear leukocytes in cattle. In: *Proceedings of the 11th International Congress on Animal Reproduction and AI,* Vol. 4. Dublin, pp. 528-530.

Wiegerinck W, Setkus A, Buda V, Borg-Karlson AK, Mozuraitis R and de Gee A. 2011. BOVINOSE: pheromone-based sensor system for detecting estrus in dairy cows. *Proced Comput Sci* 7:340-342.

Williams EJ and Sheldon IM. 2003. Clinical evaluation of postpartum vaginal mucus reflects uterine bacterial infection. *Reproduction*, Abstr Ser 30:91.

Williams EJ, Fischer DP, Noakes DE, England GC, Rycroft A, Dobson H and Sheldon IM. 2007. The relationship between uterine pathogen growth density and ovarian function in the postpartum dairy cow. *Theriogenology* 68:549-559.

Williams EJ, Fischer DP, Pfeiffer DU, England GC, Noakes DE, Dobson H and Sheldon IM. 2005. Clinical evaluation of postpartum vaginal mucus reflects uterine bacterial infection and the immune response in cattle. *Theriogenology* 63:102-117.

Williams EJ, Sibley K, Miller AN, Lane EA, Fishwick J, Nash DM, Herath S, England GC, Dobson H and Sheldon I. 2008. The effect of Escherichia coli lipopolysaccharide and tumour necrosis factor alpha on ovarian function. *Am J Reprod Immunol* 60:462-473.

Williamson P, Manyua R, Martin R and Penhale WJ. 1987. Dynamics of the acute uterine response to infection, endotoxin infusion and physical manipulation of the reproductive tract in the mare. *J Reprod Fertil (Suppl.)* 35:317-325.

Wilson AD and Baietto M. 2011. Advances in electronic-nose technologies developed for biomedical applications. *Sensors (Basel)* 11:1105-1176.

Wing EJ and Remington JS. 1980. Delayed hypersensitivity and macrophage functions. In: Fudenberg HH, Stites DP, Caldwell JL Wells JV, eds. *Basic and clinical immunology*, 3rd ed., Los Altos, California: Lange Medical Publications, 129-143.

Wingfield-Digby NJ. 1987. The technique and clinical application of endometrial cytology in mares. *Equine Vet J* 10:176-180.

Wise M J. 2016. "Potential biomarkers to predict fertility in dairy cattle". Graduate Theses, Dissertations, and Problem Reports. 6963. https://researchrepository.wvu.edu/etd/6963.

Witte TS, Iwersen M, Kaufmann T, Scherpenisse P, Bergwerff AA and Heuwieser W. 2011. Determination of ceftiofur derivatives in serum, endometrial tissue, and lochia in puerperal dairy cows after subcutaneous administration of ceftiofur crystalline free acid. *J Dairy Sci* 94:284-290.

Witte TS, Melkus E, Walter I, Senge B, Schwab S, Aurich C and Heuwieser W. 2012. Effects of oral treatment with N-acetylcysteine on the viscosity of intrauterine mucus and endometrial function in estrous mares. *Theriogenology* 78:1199-1208.

Xu ZZ, Burton LJ and Macmillan KL. 1997. Reproductive performance of lactating dairy cows following estrus synchronization regimens with PGF2α and progesterone. *Theriogenology* 47:687-701.

Yadav V, Diwakar RP, Kumar P, Kumar R, Kumar P and Singh U. 2018. Antibiogram study of isolated bacteria from repeat breeding cattles in Purvanchal region of Uttar Pradesh. *J Pharmacogn Phytochem* 7:252-253.

Yah CS and Simate GS. 2015. Nanoparticles as potential new generation broad spectrum antimicrobial agents. *DARU J Pharm Sci* 23:43.

Yang F, Haile DJ, Berger FG, Herbert DC, Beveren EV and Ghio AJ. 2003. Haptoglobin reduces lung injury associated with exposure to blood. *Am J Physiol Lung Cell Mol Physiol* 284:402-409.

Yang G and Alvarez II. 1994. Ultrafast Papanicolaou stain. An alternative preparation for fine needle aspiration cytology. *Acta Cytol* 39:55-60.

Yániz JL, Santolaria P, Giribet A and López-Gatius F. 2006. Factors affecting walking activity at estrus during postpartum period and subsequent fertility in dairy cows. *Theriogenology* 66:1943-1950.

Yeaman MR and Yount NY. 2003. Mechanisms of antimicrobial peptide action and resistance. *Pharmacol Rev* 55:27-55.

Yeoman CJ, Yildirim S, Thomas SM, Durkin AS, Torralba M, Sutton G, Buhay CJ, Ding Y, Dugan-Rocha SP, Muzny DM, Qin X, Gibbs RA, Leigh SR, Stumpf R, White BA, Highlander SK, Nelson KE and Wilson BA. 2010. Comparative genomics of Gardnerella vaginalis strains reveals substantial differences in metabolic and virulence potential. *PLoS One* 26;5(8):e12411.

Yokohori N, Taira M, Kameyama S, Kanemura T, Kondo M, Tamaoki J and Nagai A. 2002. Acute form of eexogenous lipoid pneumonia caused by inhalation of liquid paraffin in a fire eater. Nihon Kokyûki Gakkai Zasshi *J Japanese Resp Soc* 40:588-593.

Yoshida R, Kitahara G and Osawa T. 2020. Intrauterine infusion of povidone-iodine: Its effect on the endometrium and subsequent fertility in postpartum dairy cows. *J Vet Med Sci* 82:926-934.

Yoshino K, Katoh N, Takahashi K. and Yuasa A. 1992. Purification of a protein from serum of cattle with hepatic lipidosis, and identification of the protein as haptoglobin. *Am J Vet Res* 53:951-956.

Young CR, Eckersall PD, Saini PK and Stanker L. 1995. Validation of immunoassays for bovine haptoglobin. *Vet Immunol. Immunopathol* 49:1-13.

Young IM and Anderson DB. 1986. First service conception rate in dairy cows treated with dinoprost tromethamine early post partum. *Vet Rec* 118:212-213.

Young IM, Anderson DB and Plenderleith RWJ. 1984. Increased conception rate in dairy cows after early post partum administration of prostaglandin F2α THAM. *Vet Rec* 115:429-431.

Young IM. 1989. Responses to dinoprost in the bovine early post partum period. *Vet Rec* 124:511-512.

Youngquist RS and Shore MD. 1997. Postpartum uterine infection. In: Youngquist, R.S. (Ed.), *Current Therapy of Large Animal Theriogenology*. W.B. Saunders Co., Philadelphia, PA, pp. 335-340.

Youngquist RS and Threlfall WR. 2006. Current Therapy in Large Animal Theriogenology. Elsevier Health Sciences.

Yuan YG, Peng Q and Gurunathan S. 2017. Effects of silver nanoparticles on multiple drug-resistant strains of *Staphylococcus aureus* and *Pseudomonas aeruginosa* from mastitis-infected goats: An alternative approach for antimicrobial therapy. *Int J Mol Sci* 18:569.

Yunis AA. 1973. Chloramphenicol-induced bone marrow suppression. *Semin Hematol* 10:225-234.

Zafracas AM. 1975. Candida infection of the genital tract in Thoroughbred Mares. *J Reprod Fertil* 23:349-351.

Zahid IA. 2004. Culture and sensitivity of bacterial growth from exotic cows suffering from endometritis under Pakistani conditions. *Pakistan Vet J* 24:107-108.

Zander H. 1997. Klinische Untersuchungen u¨ber die Wirksamkeit von Enzymen bzw. Enzym-/Antibiotika Kombinationen bei der Therapie von Mastitiden unterschiedlicher Genese und Manifestation. Dissertation, Berlin: Free University of Berlin, Faculty of Veterinary Medicine.

Zargham-Khan M, Muhammad G, Vmar A and Ali-Khan S. 1998. A preliminary comparison of plasma fibrinogen concentration leukocyte numbers and erythrocyte sedimentation rate and nonspecific indicators of inflammatory conditions in buffalo (Bubalus bubalis). *Vet Res Commun* 21:265-271.

Zelinski MB, Hirota NA, Keenan EJ and Stormshak F. 1980. Influence of exogenous estradiol-17 β on endometrial progesterone and estrogen receptors during the luteal phase of the ovine estrous cycle. *Biol Reprod* 23:743-751.

Zerbe E, Schuberth HJ, Hodemaker M, Gunert and Leibold W. 1996. A new model system for endometritis basic concepts and characterization of phenotypic and functional properties of bovine uterine neutrophils. *Theriogenology* 46:1339-1356.

Zerbe H, Schuberth HJ, Engelke F, Frank J, Klug E and Leibold W. 2003. Development and comparison of in vivo and in vitro models for endometritis in cows and mares. *Theriogenology* 60:209-223.

Zerobin K, Jöchle W and Steingruber Ch. 1973. Termination of pregnancy with prostaglandins E2 (PGE2) and F2α (PGF2α) in cattle. *Prostaglandins* 4:891-901.

Zhang H, Ren Y, Bao X. 2009. Simultaneous determination of (fluoro) quinolones antibacterials residues in bovine milk using ultra performance liquid chromatography–tandem mass spectrom-etry. *J Pharm Biomed Anal* 49:367-374.

Zhang XF, Shen W and Gurunathan S. 2016. Silver nanoparticle-mediated cellular responses in various cell lines: An in vitro model. *Int J Mol Sci* 17:1603.

Ziv G, Paape MJ and Dulin AM. 1983. Influence of antibiotics and intramammary antibiotic products on phagocytosis of Staphylococcus aureus by bovine leucocytes. *Am J Vet Res* 44:385-388.

Ziv G. 1980a. Review of pharmacology of antimicrobial drugs employed in veterinary obstetrics. In: *Proc 9th International Congress on Animal Reproduction.* 2:463-471.

Ziv G. 1980b. Clinical pharmacology of antibacterial drugs and their application in treating bovine metirits. In: Current Therapy in Theriogenology (Ed., D. Morrow, Saunders Co., Philadelphia). 25-45.

Zobel R and Tkalèiæ S. 2013. Efficacy of ozone and other treatment modalities for retained placenta in dairy cows. *Reprod Dom Anim* 48:121-125.

Zobel R, Tkalcic S, Stokovic I, Pipal I and Buic V. 2012. Efficacy of ozone as a novel treatment option for urovagina in dairy cows. *Reprod Dom Anim* 47:293-298.